Tube Type Dilatometers

Tube Type Dilatometers

By Joseph Valentich
WESTINGHOUSE ELECTRIC CORPORATION

INSTRUMENT SOCIETY OF AMERICA

We wish to thank the following for permission to reprint material: The American Society for Testing and Materials, 1916 Race Street, Philadelphia PA 19103; William A. Plummer, Differential Dilatometry, A Powerful Tool, AIP Conference Proceedings No. 17 (Thermal Expansion—1973); *Journal of Materials Science* 14 1979, Chapman & Hall Ltd., Publishers; J. Simon, Calibration of Tube Type Dilatometers, *Journal of Thermal Analysis,* Vol. II, 1977, 387–403; Thermal Expansion Coefficients for 51 Metals, *Product Engineering*, a Morgan-Grampian publication, March 1956; Aparatus for Thermal Expansion Measurement, Parts I and II, TEST, December/January 1977–1978 and February/March 1978.

Printed in the United States of America.

Instrument Society of America
67 Alexander Drive
P.O. Box 12277
Research Triangle Park
North Carolina 27709
U.S.A.

LIBRARY OF CONGRESS
CATALOG CARD NO.: 80-82116

Valentich, Joseph
Tube Type Dilatometers.

Research Triangle Park, N.C.: Instrument
Society of America

244 p.

8006 800515

ISBN 0-87665-468-X

Book design and production by
Publishers Creative Services
Printing and binding by
Halliday Lithographic Corporation

To my daughter Susan
and my son Joseph
in recognition of
their academic accomplishments

Contents

List of Tables

List of Figures

Preface

The author has spent about twenty years in thermal expansion testing at the Westinghouse Electric Corporation and is a member of the ASTM Task Group E37.05.01A on dilatometry. During this time thermal expansion testing has been carried out daily, attesting to the importance of this thermal property. The materials tested have ranged from experimental alloys to those used in products made by Westinghouse. The products include such items as turbine generators, steam turbines, water wheel generators, circuit breakers, nuclear power plants, electrical contacts, current reactors, etc.

Because of the great number of tests that had to be conducted, the most economical test procedures consistent with high accuracy had to be established. For this reason the continuous plot of thermal expansion versus temperature employing a tube type dilatometer was established.

Some books in the field of thermal expansion have been available since 1970. These are publications of the International Thermal Expansion Symposia which were started in 1968. These thermal expansion symposia are held under the auspices of an independent governing board. The purposes of the symposia are to provide a forum for all colleagues interested in fundamental or applied aspects of thermal expansion in order to cover recent developments in this field. The proceedings of these symposia, with the exception of the first held in 1968, have been published.

There are a number of different techniques used to measure thermal expansion, and each technique has its limitations. These include the tube type dilatometer, x-ray diffraction method, optical systems, resonance techniques, interferometry, the capacitance method, the resistance strain gage method, etc. In this book primary emphasis is given to tube type, sometimes called pushrod type dilatometers. It is not intended to negate the others; it so happens that the author uses the tube type dilatometers exclusively and therefore has first-hand knowledge about them. The tube type dilatometer permits the investigator to test from −271°C to 2200°C as reported here. These are not the temperature limits of tube type dilatometers but the limits of experience by the author. For information on dilatometers other than the tube type, the reader will have to go to the technical publications.

With the advent of the International Thermal Expansion Symposia, which are held every two years, there is a better understanding of the problems in thermal expansion testing among investigators which should tend to standardize the procedure used. The use of secondary standards supplied by the U.S. Bureau of Standards should help to this end. The ASTM standards, which were needed many years sooner, should provide a common ground for testing. Perhaps this book will be helpful to the same end.

This book is made up of eleven chapters and one appendix, and each chapter is complete by itself. No chapter is a prerequisite for another. The equipment in each chapter is fully described and evaluated. As a result, some duplication exists. This presentation was adopted to permit the reader to obtain all the necessary information about a dilatometer for a specific temperature range in one chapter since dilatometers are generally categorized for a temperature range.

Chapter 1 is an introduction to dilatometers and the properties of vitreous silica. The information about vitreous silica was obtained from reference 14 in Chapter 1. Chapter 2 through Chapter 5 fully describe tube type dilatometers for various temperature ranges within the total range −195°C to 2200°C. A specially designed dilatometer for the measurement of the thermal expansion of plates is described in Chapter 6. Chapter 7 is W. A. Plummer's paper on Differential Dilatometry. He is considered one of the foremost authorities in differential dilatometry. The information in Chapter 8 was taken from the manufacturers' literature and is presented objectively without favoring anyone's equipment. The dilatometers described do not necessarily constitute the entire line of a manufacturer.

The effect of specimen size and preparation is discussed in Chapter 9. Some unusual specimen sizes are presented indicating the tube modifications necessary to conduct the tests. The calibration of tube type dilatometers is discussed in Chapter 10. The information presented in Chapter 10 does not depend on the knowledge of previous chapters and can be studied independently. Cryogenic dilatometers for testing from about −271°C (1.5°K) to 27°C (300°K) are presented in Chapter 11. These are dilatometers developed by other researchers and one system proposed by the author. One commercial cryogenic system suitable for a cryogenic dilatometer is presented. Finally, ASTM designation E228, "Linear Thermal Expansion of Rigid Solids with Vitreous Silica Dilatometer," is presented in the Appendix to give the reader a single source of information on tube type dilatometers.

The following people are acknowledged for their work in calibrating and conducting the tests to evaluate the dilatometers reported in this book: J. J. Irha, W. E. McMunn, and C. H. Peterson. The author would also like to acknowledge the efforts of his wife, Rose, for typing the manuscript.

CHAPTER 1

Types of Dilatometers and Vitreous Silica

Types of Dilatometers

Thermal expansion is a thermophysical property that is of interest from low temperatures approaching −273°C (0°K) to temperatures as high as 2200°C (2473°K), and no single dilatometer can be used for its measurement over the entire temperature range. Therefore, a number of tube type dilatometers had to be designed for use over some portion of the total temperature range.

It so happens that vitreous silica has a low thermal expansion and is stable over the temperature range from near absolute zero to about 1000°C and is therefore used as the tube material for thermal expansion measurements in this temperature range. Even though the vitreous silica tube can be used over this range, the heating or cooling systems required make it necessary to use several dilatometer designs. For testing from RT to 1000°C, the Nichrome wound resistance furnace is used to heat the specimen so that one dilatometer design is used for the range. For tests from RT to −195°C, the temperature of liquid nitrogen, a cold chamber replaces the Nichrome wound furnace used to 1000°C, and liquid nitrogen is used to cool the specimen; but the measuring part of the high temperature dilatometer is used. However, for measurements to near absolute zero, a completely new design of vitreous silica tube dilatometer is needed because the liquid helium used to cool the specimen requires special handling and is considerably more expensive. A dilatometer proposed by the author for testing to liquid helium temperatures is presented in this book, along with dilatometers designed by other investigators.

One might argue that the liquid helium dilatometer could be used for tests from RT to near absolute zero eliminating the need for the liquid nitrogen dilatometer. This is true, but it is the author's opinion that the cost and time consuming techniques involved with the liquid helium dilatometer prevents its use over the intermediate temperature range to the liquid nitrogen temperature. It is less expensive in time and materials to have two dilatometers for general routine testing: the liquid nitrogen dilatometer for tests to −195°C and the liquid helium dilatometer for tests to near absolute zero.

White[1], of Commonwealth Scientific and Industrial Research Organization, Division of Physics in Sydney, Australia, has constructed a copper cell dilatometer for measurements near absolute zero. He uses a special capacitance bridge to measure the thermal expansion and liquid helium to cool the specimen. The capacitance bridge is ideal for his studies because of its high resolution, in the 10^{-7} range.

It has been the writer's experience that for general thermal expansion measurements on metals, the vitreous silica tube dilatometer for measurements from RT to 1000°C and RT to −195°C is satisfactory for 90% of the thermal expansion work in the electrical industry. These two are the most popular ranges in which measurements are of interest.

The extension of the testing range from RT to 1000°C to higher temperatures is of interest to the ceramicist since he deals with ceramic materials that can resist oxidation at higher temperatures in an air atmosphere. In addition to the basic thermal expansion characteristics of ceramic materials, he is also interested in the shrinkage of ceramics at temperature, and generally this has been in the range of 1000°C to about 1600°C. To accommodate these people, an alumina tube dilatometer was designed. Basically it has about the same design as the vitreous silica tube dilatometer except that the measuring transducer must be cooled because of the conduction of heat along the tube and rod to the transducer. This is not a problem for measurements to 1000°C. The alumina tube has one other drawback and that is its comparatively high thermal expansion. This means that greater care must be exercised with the alumina tube dilatometer than with the vitreous silica dilatometer because larger errors can be expected impairing the accuracy.

Alumina was chosen as the tube material because experience has shown that it is more stable than other ceramics and it has been accepted as the ideal tube material for tests above 1000°C in air. The alumina tube dilatometer was limited for use to 1600°C principally because a platinum wound resistance furnace is convenient to heat the specimen economically and its maximum limit is 1600°C. Actually the alumina tube can be used to higher temperatures.

The same measuring apparatus used for the vitreous silica dilatometer can be used for the alumina tube dilatometer. As a matter of fact, one could use the same basic dilatometer for measurements from −195°C to 1600°C by just exchanging furnaces, cooling chambers, and tubes and rods. This is not recommended because the cost involved in making two dilatometers: (1) a vitreous silica tube dilatometer for measurements from −195°C to 1000°C and (2) an alumina tube dilatometer from RT to 1600°C, is more economical than switching around. Greater accuracy is also attainable by using two dilatometers.

Up to this point the author has discussed equipment for the measurement of thermal expansion from near absolute zero to 1600°C in air. With the advent of space travel, the measurement of thermal expansion above 1600°C is of interest because space vehicles encounter temperatures above 1600°C in vacuum. For this reason, a tantalum tube vacuum dilatometer was designed for measurements to 2200°C. Tantalum was chosen as the tube material because it was available at a reasonable cost and it has proven to be thermally stable to obtain accurate results. Tungsten would probably be a better tube material and it has been available at a reasonable cost during the past ten years. Tantalum tubes like alumina tubes have a comparatively high thermal expansion; and like the alumina tube dilatometer the tantalum tube dilatometer requires greater care to obtain accurate thermal expansion data. Since the dilatometer is used to 2200°C, it is water cooled; the specimen is heated with a shielded tantalum resistance heater.

W. A. Plummer[2], of the Corning Glass Works, Corning, New York, has reported the design of a vitreous silica tube differential dilatometer which can be used to a maximum temperature of 1000°C. With the differential dilatometer, the thermal expansion of a specimen is measured relative to a reference material. With the non-differential tube type dilatometers the method is a comparison method in which the thermal expansion of the specimen is compared to the tube material. With the differential dilatometer the reference material is tested along with the specimen: both are held in a vitreous silica tube, and the measured quantity is the difference in thermal expansion of the specimen and the reference material. The primary advantage of the method is that with a single test one can determine the difference in expansion between two materials with greater precision and accuracy, according to Plummer.

Peter S. Gaal, of the Anter Laboratories, Pittsburgh, Pennsylvania, has reported a special dilatometer which he used at the Westinghouse Astronuclear Laboratory. The system is capable of measuring the thermal expansion of ten specimens concurrently. The specimens are mounted upright on a pedestal between two heating elements and are visible through viewing ports from the front and back of the furnace. The target edges of the specimens are easily distinguishable when viewed through a telescope unit which consists of two 50X micro telescopes. The temperature is determined with a two-color pyrometer mounted on a separate rack at the rear of the furnace and traversing synchronously with the telescopes. The thermal expansion system has a temperature range from RT to 2000°C.

The National Bureau of Standards, Washington D.C., uses a Fizeau interferometer to measure the thermal expansion of standard reference materials below and above room temperature from −195°C to 1000°C. A description of the equipment used is presented by T. A. Hahn in reference 3. Above room temperature a controlled atmosphere resistance furnace and a calibrated Pt. vs. Pt-10% Rh thermocouple are used. Below room temperature, a cryostat capable of operation with both liquid nitrogen and helium and a calibrated platinum resistance thermometer are used. The thermal expansion is calculated between equilibrium temperatures. The measurement is made over a specimen length of at least 0.2 inches.

Along with the Fizeau interferometer there are the Abbe-Pultrich and the Priest or Dental interferometers. With the Fizeau or Abbe-Pultrich interferometer the number of fringes that pass a reference point is a measure of the change in length of the specimen. When using a Priest interferometer, the change in length is indicated by the change in the number of fringes between reference lines. Three specimen designs are used, all using the principle of 3-point support. The thermal expansion of a specimen measured with the Fizeau interferometer is determined absolutely, but the expansion as measured with the Priest and Abbe-Pultrich interferometers is determined relative to the expansion of some reference material such as vitreous silica.

In addition to the three interferometers discussed above, the National Bureau of Standards has designed and built a water-cooled optical dilatometer, which was reported by B. D. Rothrock and R. K. Kirby in reference 4. This apparatus can be used for measurements to 1600°C in vacuum with a specimen 0.2 inch in diameter by 6 inches long being heated with a rhenium resistance element. An optical comparator is used to measure the change in specimen length. Temperature measurements were made using W-Re and Pt-Rh thermocouples.

It is sometimes desirable to measure the thermal expansion of a large specimen for which a special apparatus must be designed or existing equipment modified. In such a situation it is generally best to modify the existing equipment rather than design new equipment. Generally, the same measuring system can be used for the larger specimen, but the heat source, or furnace, must be modified. The author has experienced slight modifications, which are cited in the text, as well as built a new system which is covered in Chapter 6. These modifications will give the reader some insight into the variations that are possible in addition to the standard type of test in which the specimen is prepared to fit the thermal expansion apparatus.

The determination of the thermal expansion of a small sample of a material is possible with the X-ray diffraction method discussed in reference 5. The expansion which is measured by this method is that of the crystal lattice of the atom of the material. Expansion data can be obtained in different directions in anisotropic crystals of a material, when a phase change occurs or of each phase when two phases coexist. The change in specific volume due to the phase change can also be determined.

The method is to direct an X-ray beam onto a specimen. The expansion is determined only by the measurement of the diffraction angles of the reflected beam at the desired temperature and the reference temperature. The accuracy of this method in general is subject to a probable experimental error of $\pm 4\%$. The method is difficult and the equipment costly; but it is an important method because it provides data from materials not possible by other methods.

A novel method for determining thermal expansion coefficients has been devised and reported by S. F. Jacobs in reference 6. It is based on the dependence of Fabry-Perot resonances on the mirror separation. The expansion sample is formed into an etalon spacer, with highly reflecting endplates optically contacted to each end. The Fabry-Perot resonances are probed by variable radio frequency sidebands derived from a frequency stabilized 633-nm He-Ne laser. A change in sample temperature ΔT causes a change in interferometer length ΔL, which shifts the resonance frequencies by Δv. Then $\alpha = (1/\Delta T)(\Delta L/L) = (1/\Delta T)(\Delta v/v)$. α can be measured with precision limited ultimately by the stability of the stabilized laser.

The thermal expansion of metals can be measured with resistance strain gages. Measurements have been made from −195°C to room temperature and from −50°C to 175°C as reported in references 7 and 8 respectively. To use this technique, one should determine the apparent strain of the gage by attaching it to a standard material for which the thermal expansion is accurately known and subtract the known thermal response of the material from the total gage output. Apparent strain is the temperature induced output of the gage when bonded to a material having a thermal expansion coefficient of zero. When a similar gage of the same lot is then attached to a test material and cycled through the same temperature range, the apparent strain is subtracted from the total gage output to obtain the thermal expansion of the material.

Vitreous Silica

Types of Vitreous Silica

Vitreous silica is a noncrystalline material (glass) of the composition SiO_2. It is usually made either from large selected quartz crystals or from very pure quartz sand. It is available in two basic types: clear (transparent) and translucent (satiny, silky). Clear vitreous silica is made by melting large crystals of quartz, and for this reason it is often called fused quartz or quartz glass; however, it is common practice to reserve the word quartz for the commonest of the crystalline forms of silica, and to refer to the glass as vitreous silica, fused silica, or silica glass.

When vitreous silica is made from sand, the translucent or satiny type is obtained. The satiny appearance is caused by tiny air bubbles that are trapped when the material is fused and then elongated into striations when the material is drawn or worked into useful shapes. For tube type dilatometers we are entirely concerned with clear, vitreous silica, which contains no trapped bubbles of air or other gas.

Vitreous silica is marketed under the names of fused silica, quartz glass, vitreous silica, and also under a number of trademark names. Among these are Amersil, Homosil, Infrasil, and Vitreosil. Some of these names refer to special varieties having, for example, high optical transmission in a certain region of the spectrum.

Uses of Vitreous Silica

Vitreous silica is an important material in research. It is easily obtained in a state of high purity. It is chemically inert to most substances, including many strong acids; hence it is valued as a storage container and as a reaction vessel for chemicals of the highest purity. It is much used in optical work because of its wide range of transmission. It has excellent elastic properties: it transmits ultrasonic waves with very little absorption of energy (except at low temperatures); and fibers of it, being relatively immune to permanent set, make excellent electrometer suspensions and delicate spring balances. Vitreous silica has a very small coefficient of thermal expansion; hence it is highly resistant to thermal shock. As a refractory it can withstand temperatures of 1027°C (1300°K) for long periods, and higher temperatures for short periods. It is an excellent electrical insulator.

Characterization of Vitreous Silica

Vitreous silica consists of a network of tetrahedrons of SiO_2, loosely packed. Each silicon atom is tetrahedrally surrounded by 4 oxygen atoms, and each oxygen is shared by 2 silicons; but long-range order is lacking. In a crystal all corresponding Si-O-Si bond angles are equal, but in the glass this angle varies somewhat; also, according to Warren and Biscoe[9], the orientation of adjacent tetrahedra about these bonds is random in the glass. The X-ray analysis of vitreous silica shows the short-range, near-neighbor order that is characteristic of liquids and glasses but no order extending over a distance as large as 3 or 4 times a unit-cell dimension.

The density of vitreous silica has a slight dependence on the rate at which the material is cooled from the molten state. Sosman[10] selected 2.203 g cm^{-3} as the best value for the density of clear vitreous silica and stated that different samples were likely to differ in density by as much as 0.001 percent. Subsequent work gives little reason for differing from Sosman's conclusions. Crawford and Cohen[11] observed both the density and the thermal conductivity of vitreous silica before and after neutron bombardment. Using a hydrostatic method, they found the density at room temperature to be 2.2002 g cm^{-3} before bombardment. This value was increased by 2.7% during bombardment, but subsequent annealing substantially restored the original value. The lack of long-range order in vitreous silica is associated with an open molecular structure with numerous vacant sites. These vacant sites reduce the density of vitreous silica to a value about 17% less than that of crystalline quartz.

Vitreous silica does not have a sharp melting point. The crystalline form of silica that is stable at the highest temperature is cristobalite[10]; it melts at about 1710°C (1983°K) and becomes a liquid of high viscosity.

Devitrification of Vitreous Silica

The high-temperature limit of usefulness of vitreous silica is generally the temperature at which devitrification (crystallization) occurs rather than that at which the material begins to soften or melt. Below about 1000°C (1273°K) it shows no tendency to crystallize; but at 1100°C and above, minute crystals gradually form. The crystals become visible when the material is cooled, as they undergo a transformation that gives them a white, chalky appearance. Devitrification is accompanied by a decrease in mechanical strength and an increase in tendency to shatter. It proceeds more rapidly as the temperature is raised; it may be complete in 4 to 8 hours, or it may be a matter of days. Devitrification begins at a surface; it is accelerated by water vapor, by alkalis, and by many other substances. It is retarded when the amount of oxygen in the vitreous silica is slightly less than stoichiometric[12]. High-quality transparent vitreous silica, in the absence of catalysts, has been held at 1300°C (1573°K) for several hours with only a slight cloudiness resulting.

The product of devitrification is usually cristobalite, even at temperatures where cristobalite is not the most stable form of silica.

Effects of OH

Although clear vitreous silica made by melting crystals is ordinarily very pure, there are usually traces of metals in it. For some optical work, these traces of metals must be avoided; for this purpose silica is made by reacting $SiCl_4$ with water and depositing the SiO_2 in the glassy state. Vitreous silica made from $SiCl_4$ is metal-free, but it contains a relatively large amount of OH, which comes from the water formed in the reaction[13]. This OH can amount to more than 0.001 % by weight, whereas vitreous silica made by melting natural crystals contains only about 3 parts of OH per million. This difference in OH content causes small changes in density, thermal expansion, index of refraction, and other properties. But perhaps its greatest importance is in spectroscopic experiments that require transmission of the OH frequencies; the presence of OH in the glass causes these frequencies to be absorbed.

References

1. G. K. White, "Measurement of Thermal Expansion at Low Temperatures," Commonwealth Scientific and Industrial Research Organization, Sydney, Australia.
2. W. A. Plummer, *AIP Conference Proceedings No. 17, Thermal Expansion, 1973*. R. E. Taylor and G. L. Denman, eds., American Institute of Physics, New York, 1974, p. 147.
3. T. A. Hahn, "Thermal Expansion of Copper from 20 to 800°K—Standard Reference Material 736," *Journal of Applied Physics*, Vol. 41, No. 13, Dec. 1970, pp. 5096-5101.
4. B. D. Rothrock and R. K. Kirby, "An Apparatus for Measuring Thermal Expansion at Elevated Temperatures," *Journal of Research of the National Bureau of Standards*, Vol. 710, No. 2, Apr.-June 1967.
5. F. A. Mauer, and L. M. Bolz, "Measurement of Thermal Expansion of Cermet Components at High Temperatures: X-Ray Diffraction," *WADC TR* 55-473, 1965, AD 95329.
6. S. F. Jacobs, J. N. Bradford, and J. W. Berthold III, "Ultraprecise Measurement of Thermal Coefficients of Expansion," *Applied Optics* Vol. 9, No. 11, Nov. 1970, p. 2477.
7. T. E. Finke and T. G. Heberling, "The Determination of Thermal Expansion Characteristics of Metals Using Strain Gages," *Experimental Mechanics*, Apr. 1978, p. 155.

8. M. W. Moore and K. F. Kesterson, "Measuring the Thermal Expansion of Solids with Strain Gages," *Journal of Testing and Evaluation*, Vol. 6, No. 2, March 1978, pp. 98-102.
9. B. E. Warren and J. Biscoe, "The Structure of Glass by X-Ray Diffraction Studies," *Journal of the American Ceramic Society* 21, 1938, pp. 49-54.
10. R. B. Sosman, "The Properties of Silica," The Chemical Catalog Co., Reinhold Pub. Corp., New York, 1927.
11. J. H. Crawford Jr. and A. F. Cohen, "The Effect of Fast Neutron Bombardment on the Thermal Conductivity of Silica Glass at Low Temperature," *Bulletin de L'Institut International du Froid, Annexe* 1958-1 (1958), pp. 165-172.
12. F. E. Wagstaff, S. D. Brown, and I. B. Cutler, "The Influence of H_2O and O_2 Atmospheres on the Crystallization of Vitreous Silica," *Physics and Chemistry of Glasses* Vol. 5, 1964, pp. 76-81.
13. G. Hetherington and K. H. Jack, "Water in Vitreous Silica, Part 1: Influence of Water Content on the Properties of Vitreous Silica," *Physics and Chemistry of Glasses* Vol. 3, (1962), pp. 129-133.

CHAPTER 2

Dilatometer for Measurements from 25°C to 100°C in Air

Introduction

Plastics are being used more and more in packaging, automobiles, circuit breakers, circuit board panels, etc. The thermal expansion of the plastic is an important parameter to about 100°C in these applications. If one does not have a dilatometer for thermal expansion measurements, an inexpensive one can be built for this temperature range. Most of the items necessary are generally available in the laboratory, with the exception of the vitreous silica tube and rod and perhaps the measuring gage. The gage advocated for this system is the Tuckerman gage, or an LVDT. The Tuckerman gage is described later in this chapter and in greater detail in references 1 and 2. However, a dial gage, normally available in most laboratories, can also be used with some sacrifice in accuracy. The system is easily calibrated using a specimen made from certified OFHC copper, which is readily available. The system also lends itself to testing several specimens simultaneously. The author has tested four specimens simultaneously.

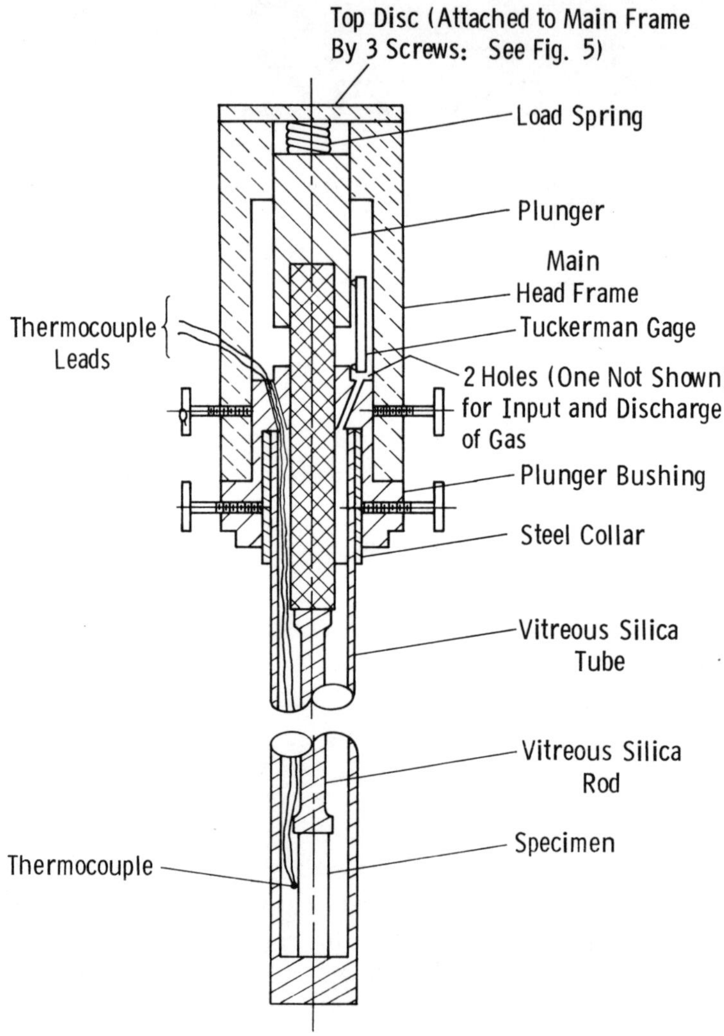

FIGURE 2–1.—Vitreous silica tube assembly with Tuckerman gage head.

Dilatometer Elements

The dilatometer consists of a fused silica tube and rod with the thermal expansion measuring gage attached to the tube, the temperature measuring elements, and the heat source with its controls. The tube assembly, using a Tuckerman gage to measure the expansion, is shown in Figure 2-1. A typical setup for measuring the ther-

FIGURE 2–2.—Typical setup for measuring the thermal expansion of four specimens.

mal expansion of four specimens simultaneously is shown in Figure 2-2.

The water bath temperature controller is not shown in Figure 2-2. A disassembled fused silica tube, including the Invar head to mount the measuring gage, the Tuckerman gage, is shown in Figure 2-3. A tube system with a dial gage and an LVDT in place of the Tuckerman gage to measure the thermal expansion is shown in Figure 2-4. The detailed dimensions of the invar head to mount the Tuckerman gage are given in Figure 2-5.

Vitreous Silica Tube

The nominal size tube has an inside diameter of ½ inch, outside diameter of ¾ inch, and a length of about 20 inches. The inside

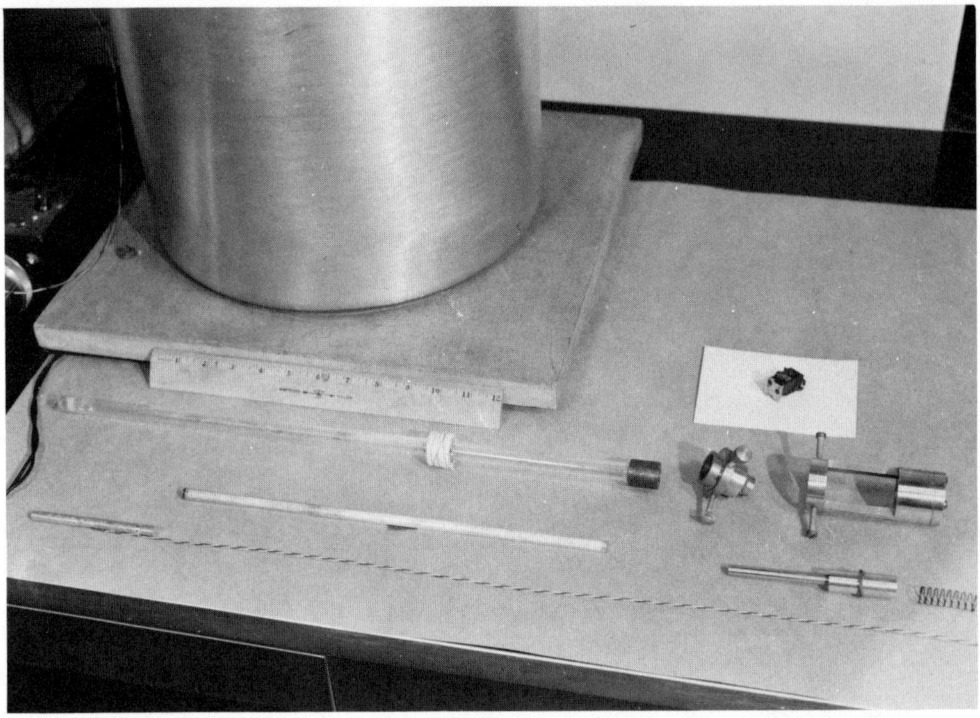

FIGURE 2–3.—Elements of vitreous silica tube and Tuckerman gage head assembly.

bottom of the tube should be flat and perpendicular to the axis of the tube so that when the specimen is placed in the tube, it is perpendicular. An invar ring approximately one inch long with a wall thickness of about ⅛ inch is fixed to the open end of the tube with Duco or an equivalent cement. The invar head, to which the Tuckerman gage or dial gage is fixed, is held to the Invar ring with two thumb screws (Figures 2-1 and 2-4). An asbestos heat guard is wrapped around the tube in Figure 2-3. This tube is used with a resistance wire heated furnace, and the asbestos guard fits on the furnace opening radially from the outside diameter of the tube. It prevents heat from being radiated upward to the measuring gage. If a water bath is used for heating, the asbestos heat guard can be used as a buffer between the clamp and the tube as shown in Figure 2-2.

Vitreous Silica Rod

The rod is nominally ¼ inch in diameter and is shorter than the tube by the length of the specimen, usually 2 inches plus the dis-

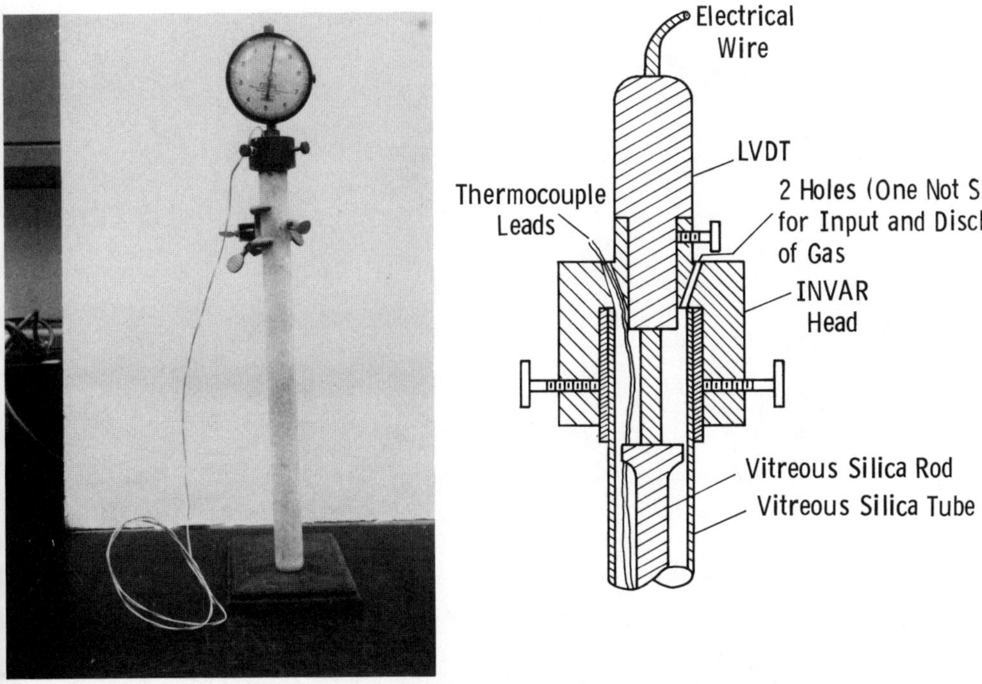

FIGURE 2–4.—Vitreous silica tube with dial gage and LVDT to measure thermal expansion.

tance the bottom of the plunger protrudes into the open end of the tube. The ends of the rod should be dumbbell shaped, as shown in Figure 2-1 with about a 1/32 inch clearance on the inside of the tube, and the ends must be flat and parallel so that the rod will fully contact the end of the specimen at the bottom of the tube and the plunger at the open end of the tube. A small slot parallel to the axis of the rod may have to be cut into the dumbbell at the open end of the tube to permit the passage of a thermocouple from the specimen to the measuring potentiometer outside the tube.

The dimensions of a typical head assembly are shown in Figure 2-5. For best results, the head should be made from Invar, which has a low expansion and will contribute a minimum error to the expansion measurement, if it becomes heated during the test. The head will expand from changing room temperature or from the heat of the water bath. The expansion of the head will be sensed by the Tuckerman gage, which measures to 4 x 10^{-6} inches/inch or by the LVDT. This will introduce an error in the expansion measurement. The effect of head expansion is small, and the dial gage, when used,

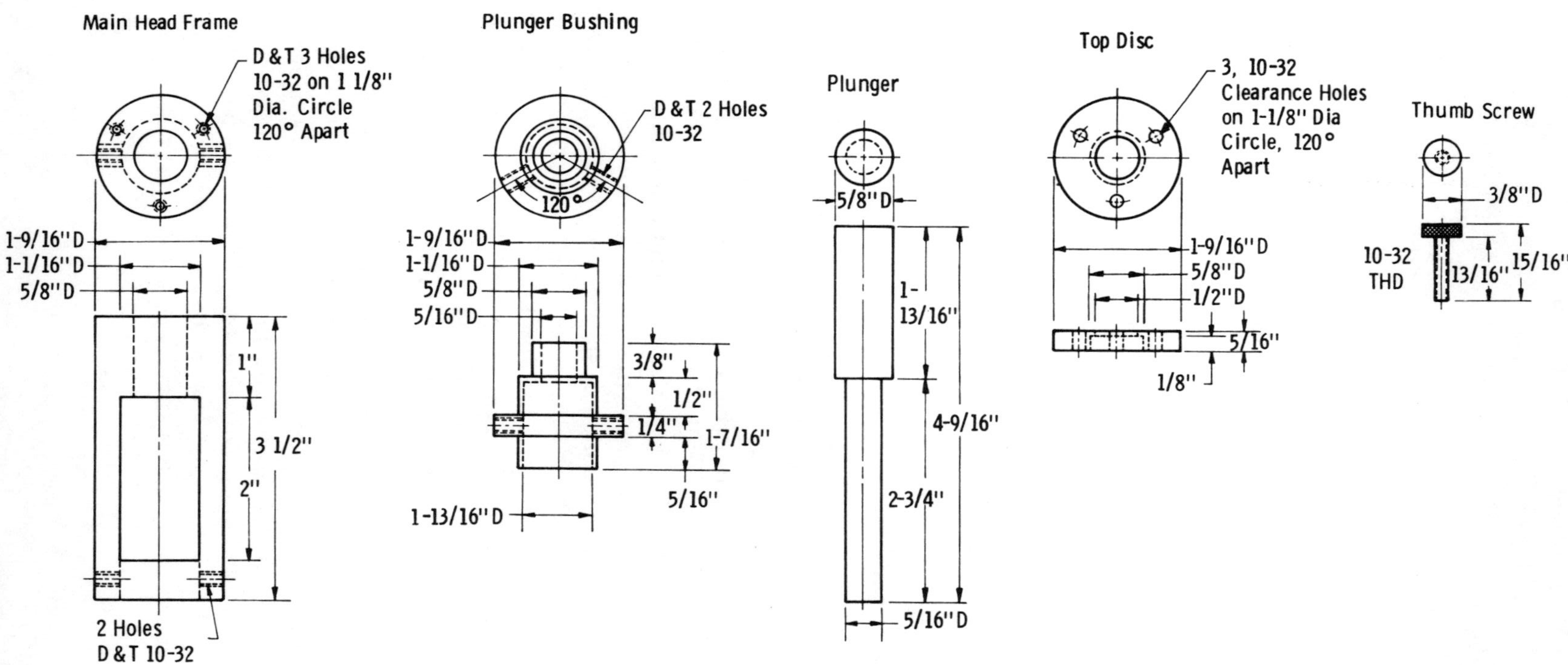

Note: 1. Plunger must slide freely in main head frame and plunger bushing
2. All parts made from INVAR

FIGURE 2–5.—Dimensions of Invar head elements.

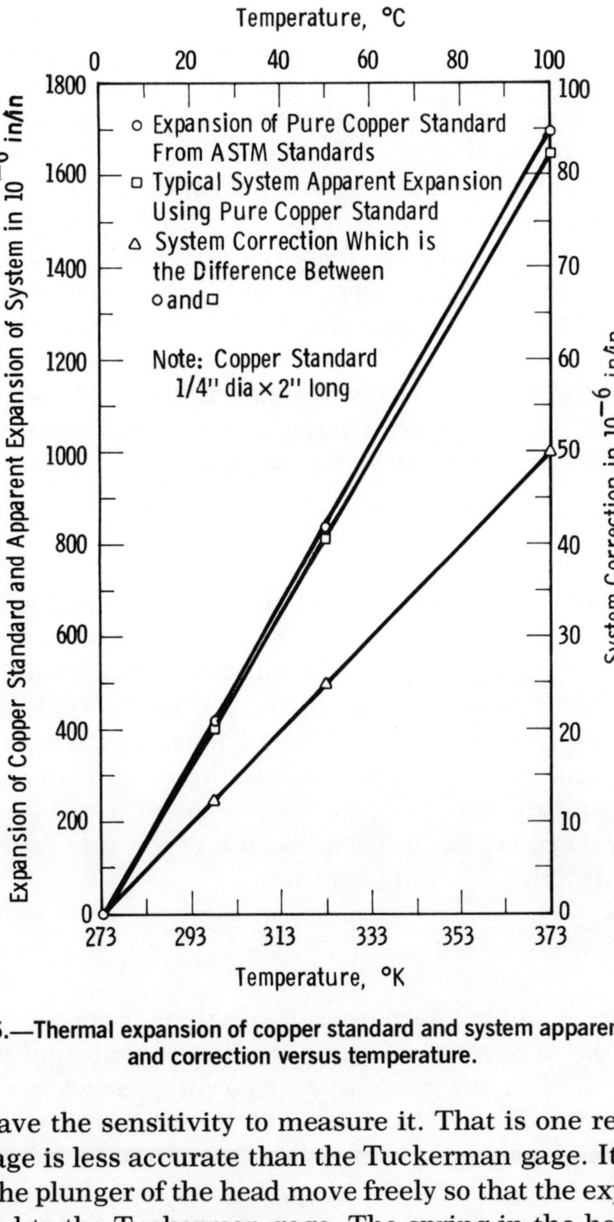

FIGURE 2–6.—Thermal expansion of copper standard and system apparent expansion and correction versus temperature.

will not have the sensitivity to measure it. That is one reason why the dial gage is less accurate than the Tuckerman gage. It is important that the plunger of the head move freely so that the expansion is transmitted to the Tuckerman gage. The spring in the head should be large enough to produce a downward force of about three ounces when it is in place during a test. When a dial gage is used, the spring action in the dial gage replaces the head spring although it is generally not strong enough to give a three-ounce downward load. When a thermal expansion measurement is made with a dial gage, it will be necessary to tap or vibrate the gage lightly with the eraser on the end of a pencil in lieu of the three-ounce downward pressure to make certain that the gage is not hanging up. This action is especially necessary in making readings on the cooling cycle when the gage must follow the contraction of a specimen.

The LVDT is designed with a return spring to keep the stem in contact with the vitreous silica rod. It will follow the movement of the rod during expansion and contraction without the need of the vibration or tapping recommended for the dial gage. Most manufacturers of dilatometers use the LVDT to measure the thermal expansion.

In addition to the thermocouple hole in the head, two other holes are provided to introduce an inert gas atmosphere into the tube. One hole is for input and the other for discharge of the gas.

Water Bath

The water bath can be a 5-gallon paint can filled with water to within about four inches of the top. A stirrer must be placed in the water to keep it agitated. The temperature variation in a stirred water bath can be kept to less than 0.5°C throughout the bath. An immersion type heater with a bimetallic temperature controller or an equivalent can be used to maintain the water bath temperature. If a temperature controller is not available, a voltage controller wired to the heaters can be used to regulate power to the heaters and hence control the bath temperature.

Temperature Measurement

A chromel-alumel thermocouple wired to the specimen is used to measure the specimen temperature. It is brought out of the tube through a hole in the Tuckerman gage invar head or the head to which the dial gage is mounted. The thermocouple is read with a precision L&N potentiometer, or an equivalent potentiometer. Any other thermocouple will also be satisfactory. An immersion mercury thermometer can be used to monitor the bath temperature.

Tuckerman Optical Gage, LVDT, and Dial Gage

A Tuckerman gage with a one-inch gage length is used to measure the thermal expansion for optimum accuracy. The gage bridges the gap between the head (which is fixed to the tube) and the plunger (which is in contact with the upper end of the vitreous silica rod as shown in Figure 2-1). As the specimen expands, the rod is pushed vertically against the plunger, which also moves vertically, decreasing the size of the gap bridged by the Tuckerman gage. The Tuckerman measures this change, which is a measure of the thermal expansion. The one-inch Tuckerman can be read to 4 x 10^{-6} inches. If the specimen is two inches long, the thermal expansion can be read to 2 x 10^{-6} inches/inch. The Tuckerman gage is read with an autocollimator and is fixed to the measuring head with rubber bands.

If a dial gage is used to measure the thermal expansion, one must choose a dial gage with sufficient resolution. Gages graduated to 0.001 inch to 50 x 10^{-6} inches are available. The safest approach would be to use one with the best resolution, the one graduated to 50 x 10^{-6} inches. As mentioned above, the return spring in a dial gage is fairly weak and will not keep the gage stem in good contact with the fused silica rod. The gage face will have to be tapped lightly with a pencil to vibrate the gage and hence keep the stem in contact with the fused silica rod. This procedure is learned best through experience since all dial gages do not respond equally.

The LVDT can be any of those commercially available, provided that it has the necessary sensitivity and measuring range for the materials being tested. The use of the LVDT will require a power source and readout electronics so that this approach is more costly to build. One could automate the system using the LVDT by recording the thermal expansion and temperature of the specimen simultaneously with an *X–Y* recorder. However, this is even more costly and it is the author's aim here to present the most inexpensive system to do the job.

Specimen Size

The specimen size should be ¼ inch diameter by 2 inches long. This is the length the author advocates for all thermal expansion tests. In view of the fact that a water bath is being used to heat the specimen, the specimen can be as long as 4 inches because the variation in temperature throughout a stirred water bath is only of the order of 0.5°C. In the case of a resistance-wound furnace, the additional specimen length would cause errors, especially if the furnace temperature distribution was established for a shorter specimen.

The specimen ends should be ground flat and parallel so that the entire end areas contact the bottom of the fused silica tube and rod. If for some reason the ends cannot be ground parallel, it would be best to have a specimen with ends rounded so that the center of the ends contact the tube and rod. Ideally, the specimen should be ¼ inch in diameter, although a ¼ inch-square cross section would be a good second choice.

Inert Atmosphere

The inert gas atmosphere or some other gas atmosphere is introduced into the tube through one of the two holes in the head shown in Figure 2-1 to which the Tuckerman gage is attached. One hole serves as an input port and the other as exit port. Small diameter polyethylene tubes can be used to transmit the gas from the tank to the vitreous silica tube and duct cement, or an equivalent can be used to seal around the tube and holes in the head. Since a flowing gas atmosphere is being used, gas pressure will not be experienced and hence will not create a problem. The free end of the tube from the exit hole should be run through a small water bath so that bubbles will prove that gas is flowing, and the rate of gas flow can be regulated by the rate of bubble formation (if more exotic equipment is not available for more precise gas flow regulation).

Testing Procedure

The thermal expansion test can be conducted in two ways: (1) making thermal expansion readings as the bath (and hence the specimen) temperature continues to rise slowly or (2) stabilizing the bath (and hence the specimen) temperature at various temperatures throughout the range over which the test will be conducted. In either case, the system should be calibrated with a material of known expansion: a pure copper specimen, for example. Other calibration standards are available from the U.S. Bureau of Standards. The calibration specimen should have the same physical configuration as the test specimen. As with all other systems presented in this book it is best to make the specimen ¼ inch in diameter or ¼ inch square by 2 inches long and have the ends ground flat and parallel. If the readings are to be made while the temperature is rising, the heating rate should not be more than about 3°C/min., and the same rate should be used with the calibration specimen and the test specimen.

If the temperature stabilization technique is used to conduct the thermal expansion test, the specimen must be kept at each temperature long enough to make certain that the specimen has expanded completely. The technician should watch the measuring gage. When he determines that the gage reading has not changed over about a ten-minute period of time, he can make the reading. The bath temperature is then raised as quickly as is practical to the next temperature and the same procedure of gage watching repeated.

All tests should be conducted with increasing and decreasing temperature and the specimen length measured with a micrometer to 0.0001 inch before and after testing.

Analysis of Data

The thermal expansion data has to be corrected for the expansion of vitreous silica and any other system errors. The most accurate correction is obtained from the data generated with the copper calibration specimen or another standard. The true expansion of copper can be found in the ASTM Standards.[3] Using these values, one can ascertain the system correction and apply it to the test specimen data.

The correction curve is obtained in the following manner. A test using the copper standard is conducted as described under "Testing Procedure." The data, temperature vs. apparent thermal expansion, is plotted to show the general slope of the curve. Examination of this curve will show that the apparent thermal expansion is lower than that reported in the ASTM Standards for copper. The difference is the system correction made up of the fused silica correction and any other system errors, which must be added to the apparent thermal expansion, making the curve identical to the copper curve. These differences or correction values are the values that must be added to the test data of a specimen. A separate curve of correction values as a function of temperature should be plotted to examine the general slope of the curve and to make certain that all correction values are on the curve. If an error is made of some temperature, it will be immediately apparent from this curve. Typical curves for the copper standard, system apparent thermal expansion, and the correction values as a function of temperature are given in Figure 2-6. In this example the correction values turned out to be 0.50×10^{-6} inches/inch °C.

If less accurate data is satisfactory, the nominal thermal expansion of fused silica can be used for the correction, which would not include any system errors. The nominal values for the thermal expansion of fused silica are 0.500×10^{-6} inches/inch °C or 0.277×10^{-6} inches/inch °F. The use of these values will yield results having an accuracy of about ±3%. In most cases this accuracy is more than sufficient to solve the problem.

To obtain the most accurate data, first, several tests would have to be run to establish an accurate correction curve with the copper specimen. The specimen would then have to be tested several times to obtain an accurate apparent expansion, which is then corrected. By making several tests with the copper calibration and the test specimen, one can reduce test variability to a minimum.

Discussion

The dilatometer consists of a vitreous silica tube and rod with the thermal expansion measuring gage attached to the tube, the temperature measuring elements, and the heat source with its controls. The apparatus presented here is aimed for tests from 25°C to 100°C. There is no reason why the vitreous silica tube assembly cannot be used to test other materials to 1000°C, the generally accepted maximum temperature for vitreous silica. The water bath can be replaced with a resistance heated furnace to attain temperatures of 1000°C. For tests to 150°C, the water bath can be replaced with a silicone oil bath whose flash point is above 150°C. Or, if it is desirable to test below room temperature, the water bath can be replaced with a liquid nitrogen bath to attain temperatures of about −195°C. An acetone and dry ice mixture will give a bath temperature of about

−90°C. The water bath can be cooled with ice to reduce the starting temperature from 25°C to near 0°C, the freezing temperature of water. This makes an excellent system to measure the thermal expansion of materials through room temperature, which is not possible with most commercial systems.

Commercial dilatometers generally operate from room temperature to a higher temperature or a lower temperature but not in both directions. When data is generated in each direction separately and the two curves of thermal expansion versus temperature are plotted on the same graph, the investigator hopes for a continuous slope at room temperature. The cooled water bath permits one to circumvent this problem by permitting one to start the test below room temperature, where the errors of the starting conditions are not important; and continuous dats is obtained while heating through room temperature and above room temperature. This procedure results in one set of continuous data through room temperature having a continuous slope, providing no transformations occur in the material.

A different approach is required to make measurements below room temperature with the acetone and dry ice combination. The bath container is first filled with acetone at room temperature. Small chunks of dry (or powdered) ice are then introduced into the acetone to gradually lower the temperature. In the case of liquid nitrogen, small amounts of liquid nitrogen are added to the bath to lower the specimen temperature gradually. Finally as −195°C is approached, the entire bath container will be filled with liquid nitrogen. In general, good results can be obtained with the acetone and dry ice because the acetone bath is always in contact with the vitreous silica tube, resulting in a uniform temperature over the specimen length and the tube. The use of liquid nitrogen is more difficult because one really does not have a bath around the tube until the lowest temperature is reached. There will be a combination of liquid nitrogen and liquid nitrogen vapor present, which will result in a poor temperature distribution over the specimen length and vitreous silica tube and rod.

References

1. B. L. Wilson, "Characteristics of the Tuckerman Gage," American Society for Testing and Materials Proceedings, 1944, p. 1017.

2. L. B. Tuckerman, "Optical Strain Gages and Extensometers," American Society for Testing and Materials Proceedings, 1923, p. 602.
3. American Society for Testing and Materials Designation E228-71, "Standard Method of Test for Linear Thermal Expansion of Rigid Solids with a Vitreous Silica Dilatometer."

References Not Cited

1. J. Valentich, "New Values For Thermal Coefficients," *Product Engineering*, July 19, 1965, pp. 63-71.
2. J. Valentich, "Thermal Expansion Measurement," *Instruments and Control Systems*, Oct. 1969, pp. 91-94.
3. W. Souder, "Measurements on the Thermal Expansion of Fused Silica." Scientific Papers of the Bureau of Standards, J01.21.
4. T. A. Hahn, "Thermal Expansion of Copper from 20°K to 800°K," Standard Reference Material 736, Journal of Applied Physics 41, 5096 (1970).

CHAPTER 3

Dilatometer for Measurements from −195°C to 1000°C in Air

Introduction

The thermal expansion of materials is an important physical property to consider when temperature changes or temperature gradients are expected in a design. Without its consideration thermal stresses large enough to destroy the part may result. Undesirable deformation can occur as a result of thermal mismatches, as can failure of bonded joints. For these reasons it is important to know the thermal expansion of the materials in a design. The thermal expansion often cannot be calculated analytically; it must be measured. It is therefore necessary to have equipment to measure the thermal expansion accurately.

Most industrial designs operate in the temperature range −195°C to 1000°C. For this reason equipment was designed to measure the thermal expansion in this temperature range. In this system two runs are necessary; one from −195°C to room temperature, nominally 25°C, and one from room temperature to 1000°C. A vitreous silica tube, which is stable at temperatures up to about 1000°C, is used to hold the specimen.

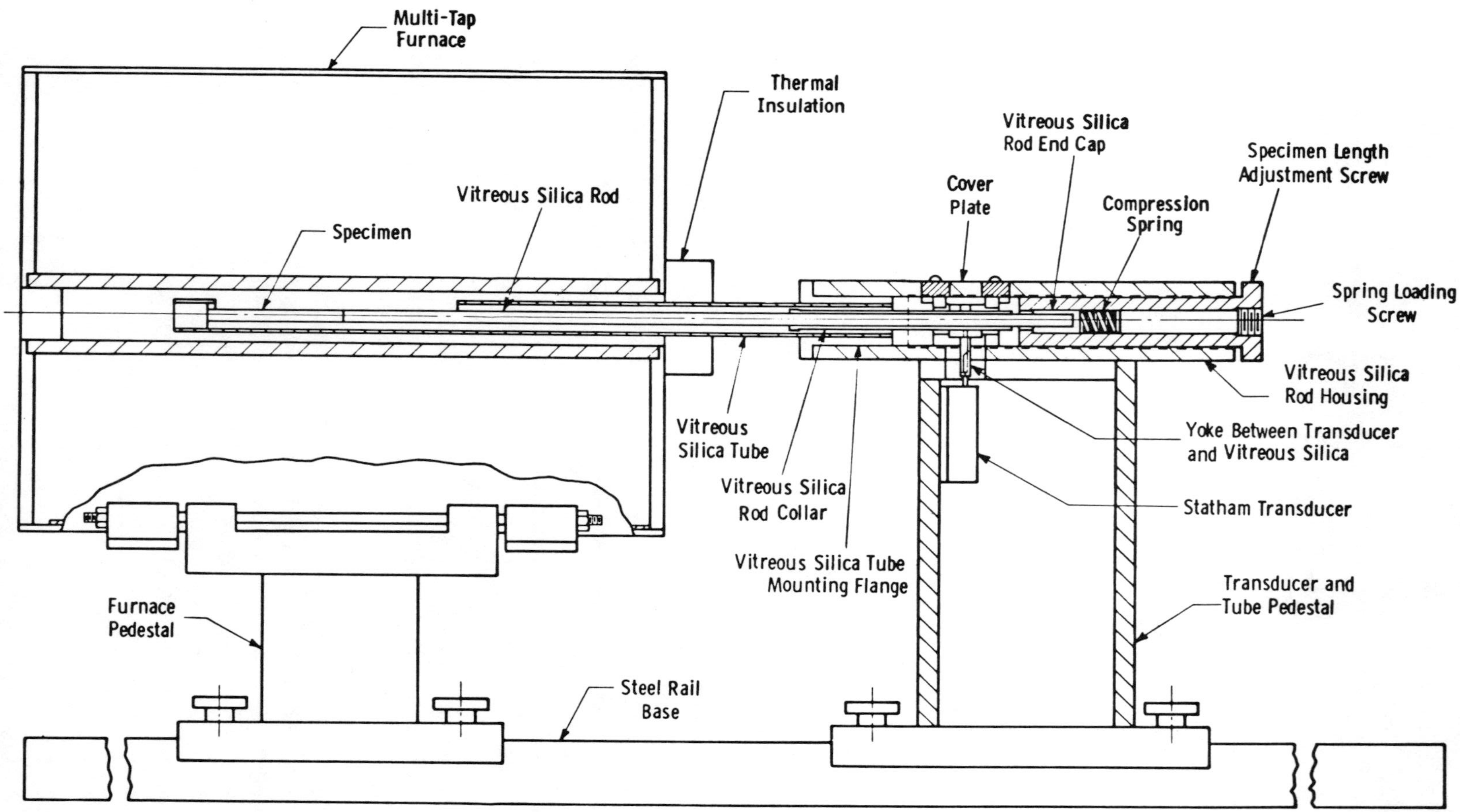

FIGURE 3–1.—Cutaway view of vitreous silica tube dilatometer.

A cooling chamber is used to cool the specimen from room temperature to −195°C; and a resistance furnace is used to heat the specimen from room temperature to 1000°C.

Dilatometer Elements

The dilatometer, shown in Figure 3-1 through Figure 3-4, is composed of a furnace or cold chamber which is used respectively to heat or cool the specimen, a vitreous silica tube and rod assembly, and a transducer system to measure the expansion of the specimen.

In this dilatometer, the specimen rests in the horizontal position

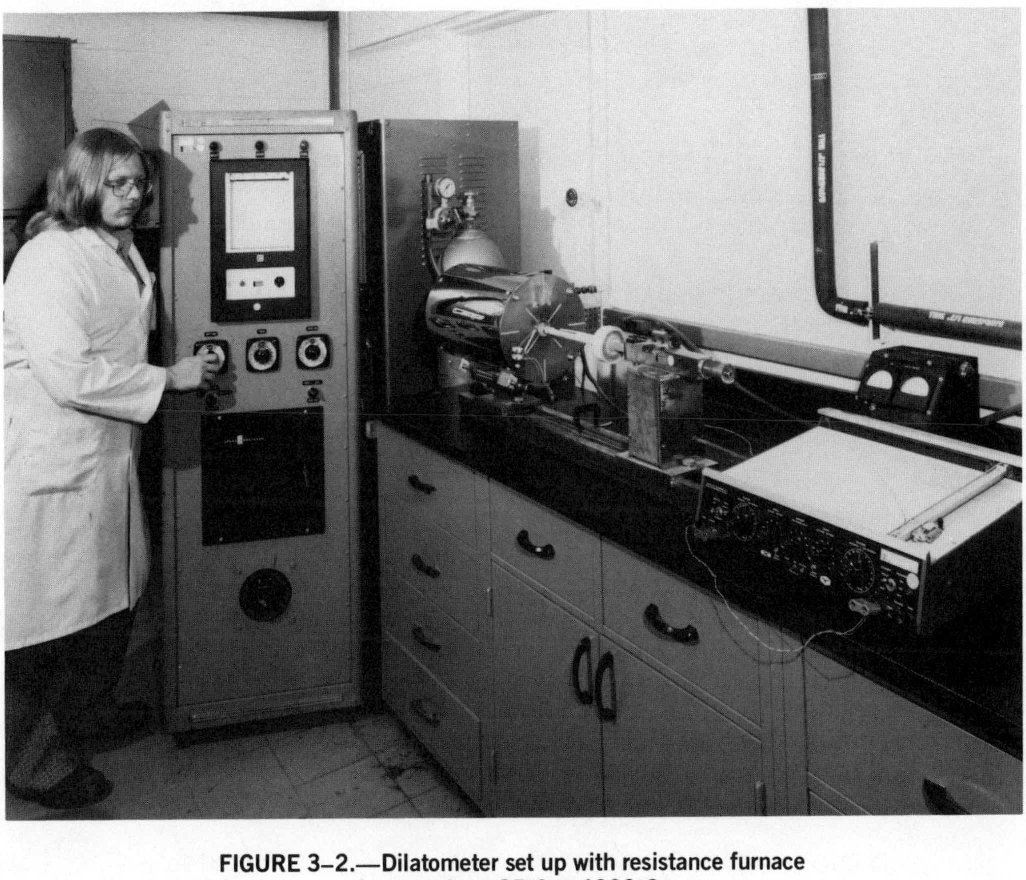

FIGURE 3–2.—Dilatometer set up with resistance furnace for tests from 25°C to 1000°C.

FIGURE 3–3.—Specimen in vitreous silica tube.

at one end of the tube as shown in Figures 3-1 to 3-4. The dilatometer is constructed on a steel rail base 14¼ inches wide by 38 inches long to which are attached two pedestals. One pedestal supports the furnace or cold chamber and the other the transducer, the tube, and the rod assembly. The furnace pedestal is fixed at one end of the base by two screws. The transducer pedestal can slide on the base at the other end to move the tube in and out of the furnace or cold chamber. The furnace and chamber are controlled with a program controller, and the temperature and thermal expansion of the specimen are recorded on an X–Y recorder.

Pedestals

The furnace and cold chamber pedestal is a heavy wall tube with a plate welded to one end and a channel welded to the other end. The

FIGURE 3–4.—Dilatometer set up with liquid nitrogen cooling chamber for tests from −195°C to 25°C.

channel rests on the steel base and is attached to it with screws. Two lugs are welded to the plate, one at each side, and each lug is made with a center hole parallel to the base. Four lugs, with center holes, are attached to the furnace; when the furnace is placed on the top plate of the pedestal, the plate lug center holes are in line with the furnace lug center holes. Two steel rods, one on each side of the plate, are passed through center holes to hold the furnace to the pedestal. Similarly, the holes in the mounting plates of the cold chamber are used to fix it to the pedestal.

The transducer and tube pedestal is a stainless steel box made from ⅜-inch thick plate. The box is mounted on a channel with screws, and the unit slides on the base, as shown in Figure 3-1 to Figure 3-4. The housing through which the vitreous silica rod passes is attached to the top plate of the pedestal. This housing also contains the specimen length adjustment screw, the spring loading screw and spring and the Invar tube mounting flange, as shown in Figure 3-1. The measuring transducer is mounted on the inside of the front plate of the pedestal, as shown in Figure 3-1, and can be removed for repairs since screws are used to hold the plates together.

Although the dilatometer was designed primarily for testing materials in air, provision is made to flow an inert gas such as argon into the tube for those cases where some protection against oxidation is desired. The transducer cover plate, shown in Figure 3-3, is made with a pipe connection through which the inert gas can be introduced. The gas flows through the vitreous silica rod housing down the tube to the specimen and out into the furnace.

Vitreous Silica Tube and Rod Assembly

The tube, which is ⅝ inch o.d. × ½ inch i.d. × 14¾ inches long, is mounted horizontally on one pedestal. One end of the tube has a cutout 4 inches long so that the specimen can be inserted; the opposite open end has an invar flange cemented to it which is used to attach the tube to the square housing above the transducer, as shown in Figure 3-1. An invar tube, 13/32 inch o.d. by 4½ inches long is cemented to the ¼ inch diameter × 18¼ inch-long rod 10¼ inches from one end, and an invar cap is cemented to the opposite end of the rod, as shown in Figure 3-5. In the tube the end of the silica rod contacts the free end of the specimen; a spring load of about three ounces is applied to the capped end to keep the rod, specimen, and bottom of the tube in good contact at all times. A yoke from the measuring transducer is fixed to the invar tube (which is cemented to the rod) with a screw as shown in Figure 3-1. As the specimen expands, it pushes the rod; the rod movement is transmitted to the measuring transducer through the yoke.

Transducer, Recorder, and Thermocouple

The transducer used to measure the thermal expansion is a Statham displacement transducer, Model G7A-5.5-660, whose dimensions are given in Figure 3-6. The transducer is excited by a d.c. power supply, and the excitation voltage level determines the sensitivity of the transducer. The transducer is mounted on the inside wall of the transducer pedestal, as shown in Figure 3-1. The excita-

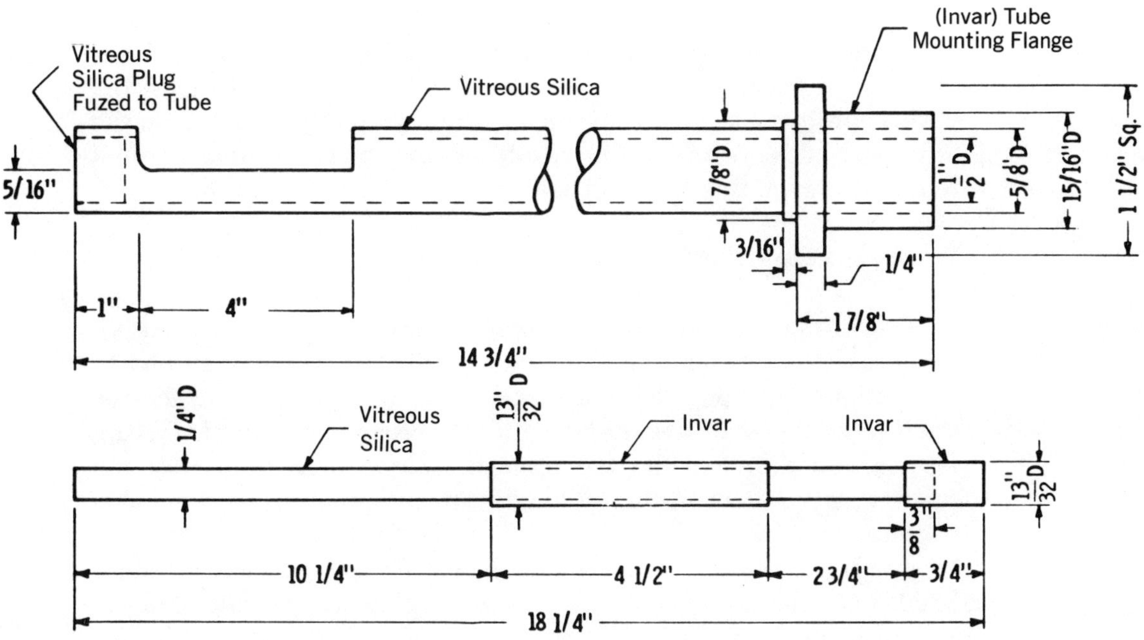

Note: All Invar Parts Cemented to Vitreous Silica Rod with Micro-Measurements, Inc. 610 Cement

FIGURE 3–5.—Vitreous silica tube and rod dimensions.

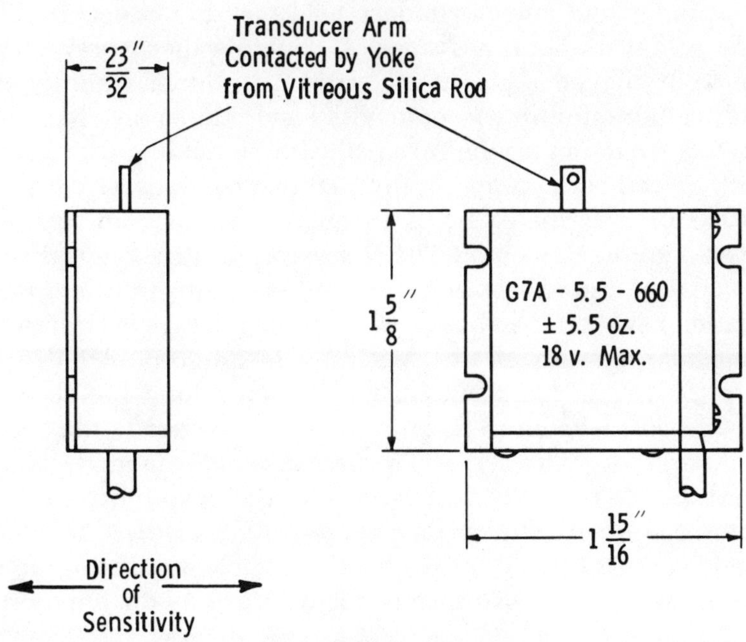

FIGURE 3–6.—Model G7A-5.5-660 Statham displacement transducer for thermal expansion measurement.

tion voltage ranges from 0 to 17V d.c. in 1 V increments. The output of the transducer is recorded on one axis of a Hewlett-Packard Model 7001A X–Y Recorder. The temperature of the specimen is measured with a chromel—alumel thermocouple; its output is recorded on the other axis of the recorder. The resultant curve is a continuous plot of temperature vs. thermal expansion.

Furnace and Controller

The furnace is a multitap ten-zone Marshall furnace with a Nichrome resistance heating element. It is 10 inches in diameter by 16 inches long with a 1-inch-diameter center hole in one end for the vitreous silica tube. The other end of the furnace is closed. During testing, the center of the specimen is located at the center of the furnace, which has an 8-inch-long uniform temperature zone. The furnace temperature is controlled by means of an L & N Controlall by which the heating rate, soak time, and cooling rate can be regulated. The specimen is normally heated and cooled at a rate of about 3°C/min.

Cooling Chamber and Controller

The liquid nitrogen cooling chamber, shown in Figure 3-7, consists of inner and outer cylinders held together concentrically by means of an aluminium spacer. A double-walled spray chamber made from copper is positioned within the inner cylinder, and a cylindrical aluminium susceptor fits inside the spray chamber immediately surrounding the fused silica tube holding the specimen. The inner wall of the copper spray chamber contains a large number of 0.020 inch-diameter holes through which liquid nitrogen is sprayed onto the susceptor. The susceptor, in turn, keeps the specimen and tube and rod assembly at a uniform temperature. Fiberfrax insulation between the chambers keeps heat losses to a minimum.

The temperature of the specimen is programmed using the L & N Controlall described earlier for controlling the furnace. A chromel—alumel control thermocouple is attached to the back end of the susceptor with a screw. The thermocouple enters the chamber through a hole in the cork pressed into the opening in the copper cooling chamber. The program controller opens and closes a solenoid-operated liquid nitrogen valve attached to the liquid nitrogen tank, which is pressurized to about 15 psi. As the program calls

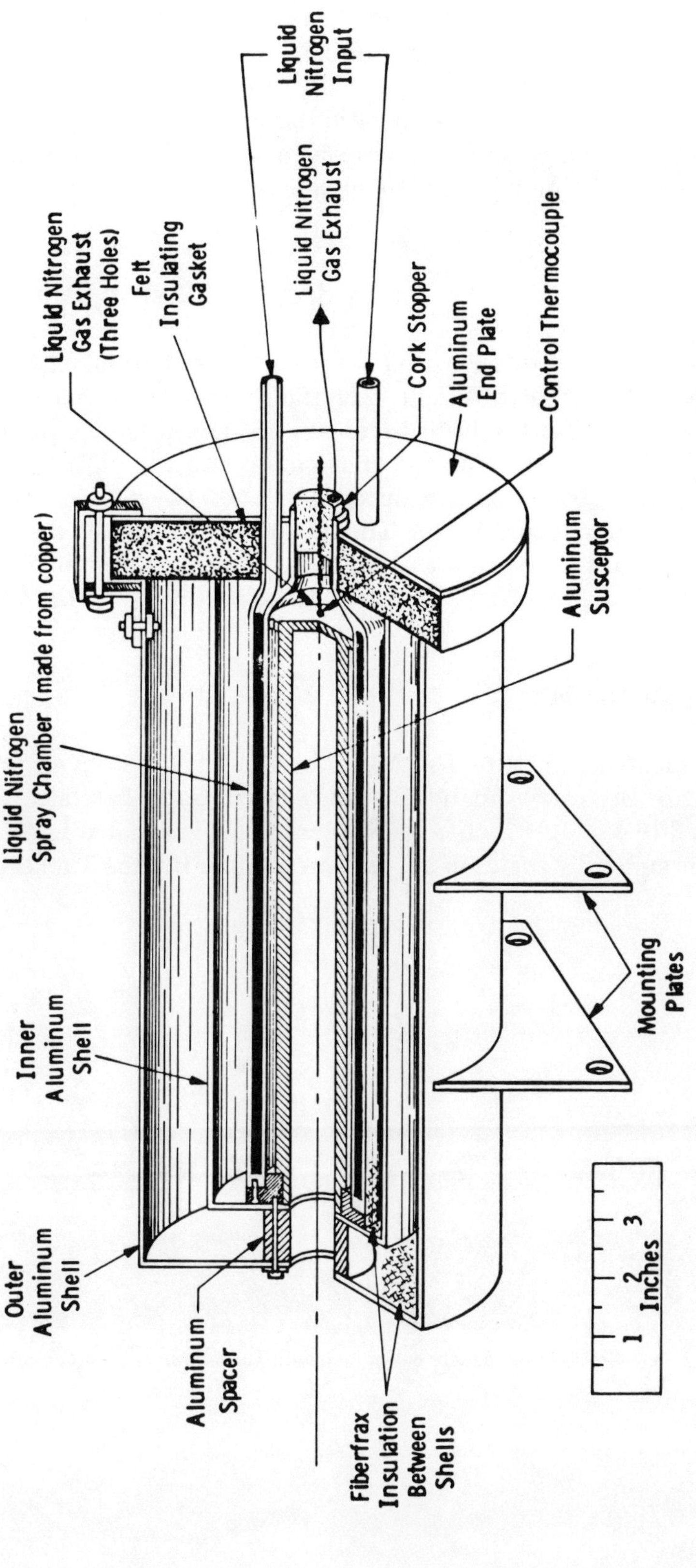

FIGURE 3–7.—Liquid nitrogen cooling chamber for thermal expansion tests.

for a lower temperature, the solenoid valve is opened allowing liquid nitrogen to flow into the copper chamber; when the control temperature is reached, as determined by the control thermocouple, the valve is closed. The rate of flow of liquid nitrogen through the solenoid valve is controlled by a handset needle valve. The solenoid valve is a Valcor Engineering Model No. 94P19C6.

Transducer and System Calibration

The Statham transducer used to measure the thermal expansion must be calibrated at different excitation voltages so that the optimum sensitivity of the transducer can be chosen for the material being tested. After the transducer is calibrated, the entire system must be calibrated to take into account any other system errors that would result in inaccurate data. The transducer is calibrated using a micrometer or a Tuckerman gage, and the overall system is calibrated with a platinum standard in accordance with ASTM specifications.

Transducer Calibration

As indicated in Figure 3-8 the calibration of the Statham transducer can be made using a Tuckerman optical strain gage graduated to 4 × 10^{-6} inches, or a micrometer graduated to 0.0001 inch. It is generally easier to use the micrometer but the Tuckerman

FIGURE 3–8.—Statham transducer calibration with Tuckerman gage or micrometer.

gage is more sensitive. The special mounting head shown in Figure 3-8 is used in conjunction with these gages.

For the transducer calibration, a compression spring and dummy specimen whose combined length is equal to the standard specimen length is placed in the tube. The spring is compressed by 25% more than the expansion it is desired to calibrate for, using the screw on the end of the mounting head if the Tuckerman gage is used or by advancing the micrometer in the direction of compression if the micrometer is used. The screw or micrometer replaces the spring loading screw of the basic dilatometer and pushes on the central plunger of the head, which contacts the silica rod and hence the compression spring through the dummy specimen. The Tuckerman gage must be read with an autocollimator. The micrometer has a graduated scale on its thimble and barrel and can be read directly.

During calibration, the output of the transducer is fed into the *X–Y* recorder, with the excitation voltage to the transducer set at a predetermined value. The gage, Tuckerman or micrometer, is then moved in increments in the direction simulating thermal expansion, and the deflection noted on the *X–Y* recorder. After the desired displacement range is covered by traversing the recorder in the expansion direction, the gage movement is reversed, simulating specimen contraction, and incremental readings are made. With this procedure the thermal expansion and thermal contraction are simulated at a fixed transducer excitation voltage and *X–Y* recorder setting. This same procedure is used for two *X–Y* recorder sensitivities at 1 V excitation voltage increments from 1 to 17 V. Six of the resulting 34 calibration curves are shown in Figure 3-9. Using these curves, one can select the best thermal expansion range for any specimen. The Statham transducer has a range of 0.032 inches, which is sufficient for most materials. If additional range is necessary, the transducer can be reset during a test.

The slopes of the curves are computed and summarized in Table 3-1 for easy daily reference in choosing the optimum thermal expansion range for a specimen. The data in Table 3-1 and the calibration curves are based on using millimeter graph paper on the *X–Y* recorder.

System Calibration

After the Statham transducer is calibrated, the entire system is calibrated from room temperature to 1000°C and from room temper-

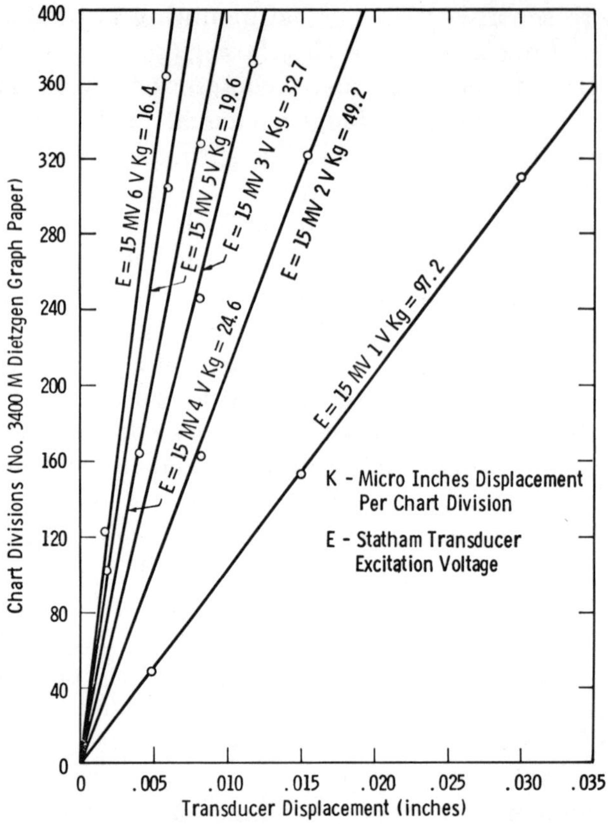

FIGURE 3–9.—Typical calibration curves for no. 934 Statham transducer.

ature to −195°C using a chemically pure platinum standard whose thermal expansion is known and documented in references 1 and 2. Chemically pure platinum is the standard recommended by ASTM. The specimen is heated and cooled at an average rate of 3°C/min. Three calibration tests are generally made; the average of the three tests is used to establish the correction curve. A typical test curve is shown in Figure 3-10 for the temperature range RT to 1000°C. A typical correction curve, uncorrected dilatometer data, and the expansion of chemically pure platinum are shown in Figure 3-11 and Figure 3-12 for the ranges 25° to 1000°C and 25° to −195°C respectively. The correction curve is obtained by determining the value that must be added to the uncorrected dilatometer data to make it identical to the expansion value of chemically pure platinum. The correction is made at increments of 100°C using room temperature

Tables 3–1
Calibration summary for Statham transducer no. 934 used with vitreous silica tube dilatometer. HP autograph *X–Y* recorder model 7001A using Dietzgen graph paper no. 340D-M (d.c. power supply no. 1). Statham transducer model no. G7A-5.5-660

X–Y RECORDER 15 mV FULL SCALE		*X–Y* RECORDER 75 mV FULL SCALE	
Transducer excitation (V) d.c.	10^{-6} in. per chart division	Transducer excitation (V) d.c.	10^{-6} in. per chart division
1	97.2	1	—
2	49.2	2	—
3	32.7	3	168.5
4	24.6	4	125.0
5	19.6	5	99.3
6	16.4	6	82.6
7	14.1	7	70.9
8	12.3	8	62.1
9	10.9	9	55.6
10	9.9	10	49.8
11	8.9	11	45.4
12	8.2	12	41.6
13	7.5	13	38.6
14	7.2	14	35.9
15	6.6	15	33.6
16	6.3	16	31.4
17	5.8	17	—

as the starting temperature on both temperature ranges. These values are then plotted as shown in Figure 3-11 and Figure 3-12 and are used in subsequent thermal expansion tests to correct the test data. By this process the correction is made for the expansion of the vitreous silica tube as well as for any other system errors that occur during the test. The reproducibility of the system is determined from the three calibration tests. It was found that the correction on heating is different from the correction on cooling in both temperature ranges as shown in Figure 3-11 and Figure 3-12. This is of no consequence as long as the test is repeatable.

Testing Procedure

After the Statham displacement transducer and the dilatometer are calibrated, specimens can be tested and the data corrected. However, frequent checks of the *X–Y* recorder must be made because its calibration can change. The *X–Y* recorder is calibrated on

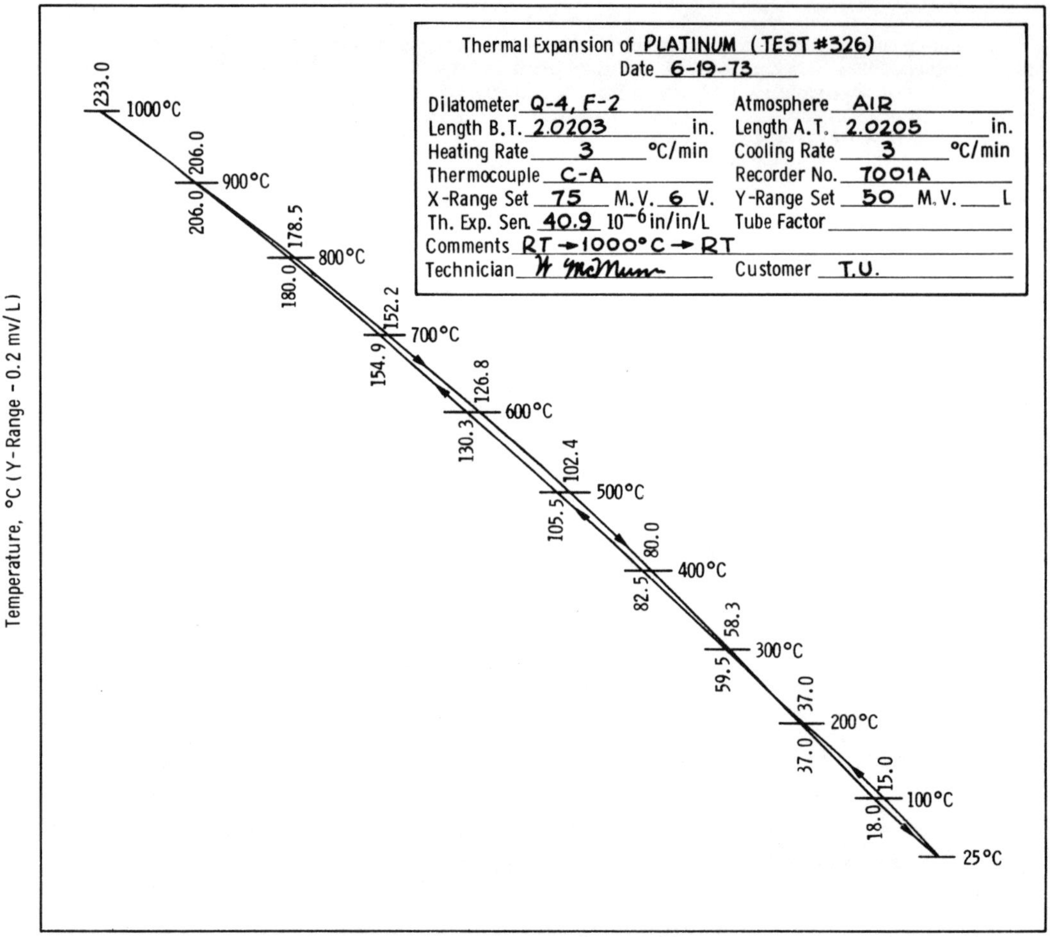

Thermal Expansion (X-Range-40.9 × 10^{-6} in/in.L.)

FIGURE 3–10.—Typical *X–Y* recorder curve of temperature versus linear thermal expansion.

both axes electrically before every test using an L & N No. 8662 precision potentiometer or an equivalent. This is accomplished by feeding millivolt signals from the potentiometer into each axis of the recorder. The *X–Y* recorder is traversed incrementally over the entire chart in both directions to make certain that the calibration is linear. The expansion axis calibration is checked on either the 15 mV or 75 mV range depending upon the expected expansion, and the temperature axis is calibrated on the 50 mV range to obtain the

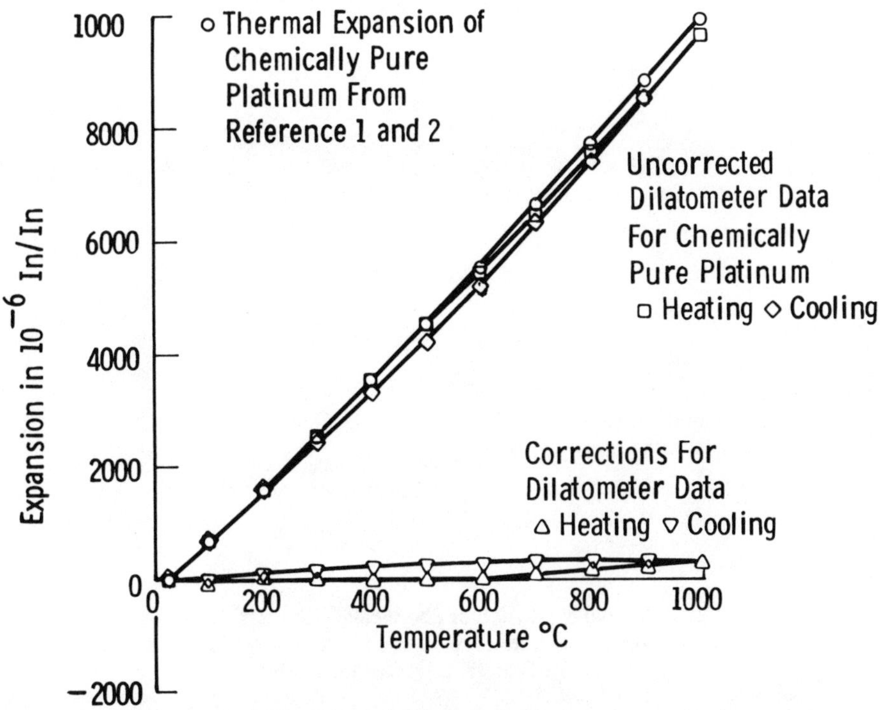

FIGURE 3–11.—Thermal expansion of chemically pure platinum, uncorrected dilatometer data for chemically pure platinum, and corrections for dilatometer data over 25°C to 1000°C temperature range.

maximum deflection across the recorder for the 25°C (room temperature) to 1000°C temperature range, which represents a change of 40.31 mV. The *X–Y* recorder is calibrated on the 10 mV range for tests from 25 to −195°C, which represents a change of 6.68 mV.

Results

The data for two typical tests conducted with the vitreous silica tube dilatometer, given in Figure 3-13 and Figure 3-14, show that the expansion results of yttrium aluminium garnet and Lucalox, a translucent alumina material, are essentially the same as those reported in the literature for each material. Additional information on these two materials is given in references 3 and 4.

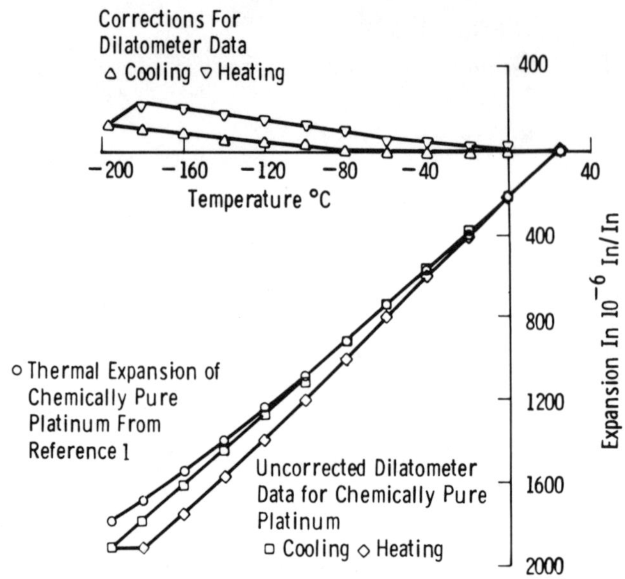

FIGURE 3–12.—Thermal expansion of chemically pure platinum, uncorrected dilatometer data for chemically pure platinum, and corrections for dilatometer data over −195°C to 25°C temperature range.

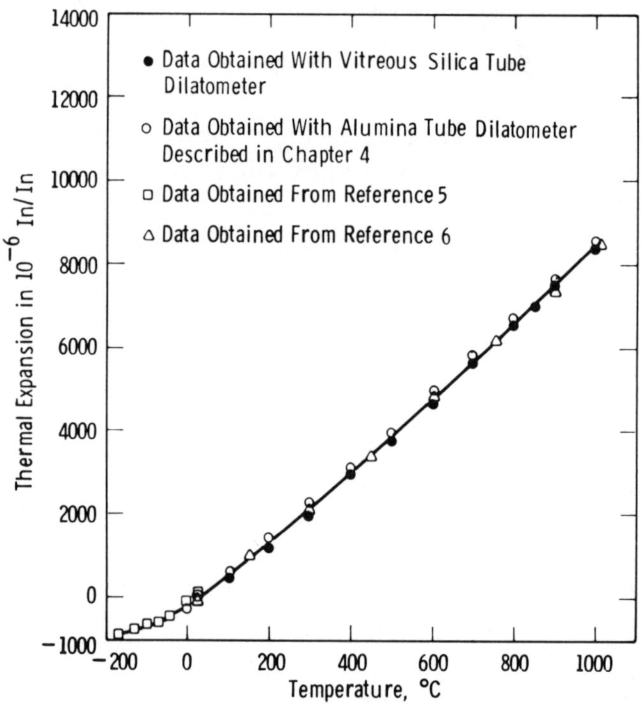

FIGURE 3–13.—Thermal expansion of yttrium aluminum garnet ($Y_3Al_5O_{12}$).

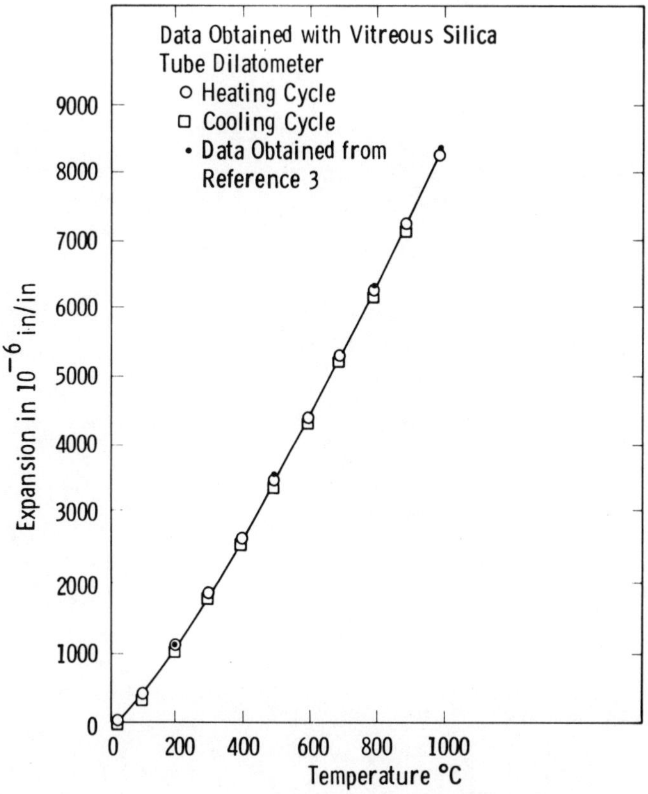

FIGURE 3–14.—Thermal expansion of Lucalox from 25°C to 1000°C in air.

References

1. American Society for Testing and Materials Designation E228-71, "Linear Thermal Expansion of Rigid Solids with a Vitreous Silica Dilatometer."
2. J. L. White, "The Suitability of Platinum, Molybdenum, Tantalum, and Tungsten as High-Temperature Length Standards." NRL Report 5159, Naval Research Lab, Washington, D.C. June 17, 1958.
3. R. R. Matheson, *Ceramic Age* June 1963, p. 54
4. T. K. Gupta and J. Valentich, *Journal of the American Ceramic Society* 54, 1971, p. 355.
5. W. J. Croft, *American Mineralogist* 50 1965, p. 1934.
6. Seymour Geller, G. P. Espinosa, and P. B. Crandall, *Journal of Applied Crystallography* 2 (Pt. 2) 1969, p. 86.

CHAPTER 4

Dilatometer for Measurements from 25°C to 1600°C in Air

Introduction

The vitreous silica tube dilatometer described in Chapter 3 and reference 1 is suitable for measuring the thermal expansion of solids from room temperature to about 1000°C, but it cannot be used above that temperature because vitreous silica devitrifies and becomes thermally unstable. Above 1000°C there are a number of oxides that can be used; of these alumina is about the most stable in air. Although alumina can be used to higher temperatures, the maximum temperature of this dilatometer was limited to 1600°C because that is the upper limit for platinum-wound furnaces, which are considered the most practical and durable furnaces for air tests in this temperature range.

The primary need for an alumina tube dilatometer is for measuring thermal expansion and for thermal stability studies of ceramic materials such as silicon nitride, silicon carbide, and boron nitride. Most structural metals are used at temperatures below 1000°C; so the vitreous silica tube dilatometer is generally suitable to measure their thermal expansion. For materials such as refractory metals and graphite, which oxidize greatly at high temperatures, a vacuum tube type dilatometer is described in Chapter 5 and reference 2. Materials that are resistant to oxidation, such as the nitrides and carbides mentioned above, are generally used in air; so it is desirable to measure their thermal expansion in air. For this reason the alumina tube dilatometer was designed and built.

Dilatometer Elements

The alumina tube dilatometer shown in Figure 4-1 is composed of a furnace to heat the specimen, an alumina tube and rod assembly, and a transducer system to measure the expansion of the specimen. The furnace is fixed to one pedestal, and the measuring system to another; and both pedestals are attached to a common steel-rail base.

In the alumina tube dilatometer the specimen rests in the horizontal position at one end of the tube as shown in Figure 4-2 and Figure 4-3. The dilatometer is constructed on a steel-rail base 14¼ inches wide by 38 inches long to which two pedestals are attached. One pedestal supports the furnace, and the other the transducer and alumina tube and rod assembly. The furnace pedestal is fixed at one end of the base by two screws. The transducer pedestal can slide on the base at the other end to move the alumina tube in and out of the furnace. The furnace temperature is controlled with a motor-controlled variac, and the temperature and thermal expansion of the specimen are recorded on an X–Y recorder. An air-cooled jacket is mounted around the alumina tube adjacent to the measuring transducer to keep it at room temperature.

Pedestals

The furnace pedestal shown in Figure 4-1, Figure 4-2, and Figure 4-3 is a heavy wall tube with a plate welded symmetrically on one end and a channel welded symmetrically on the other end. The channel rests on the steel base, and the furnace is attached to the top plate.

The transducer and alumina tube pedestal is a stainless steel box

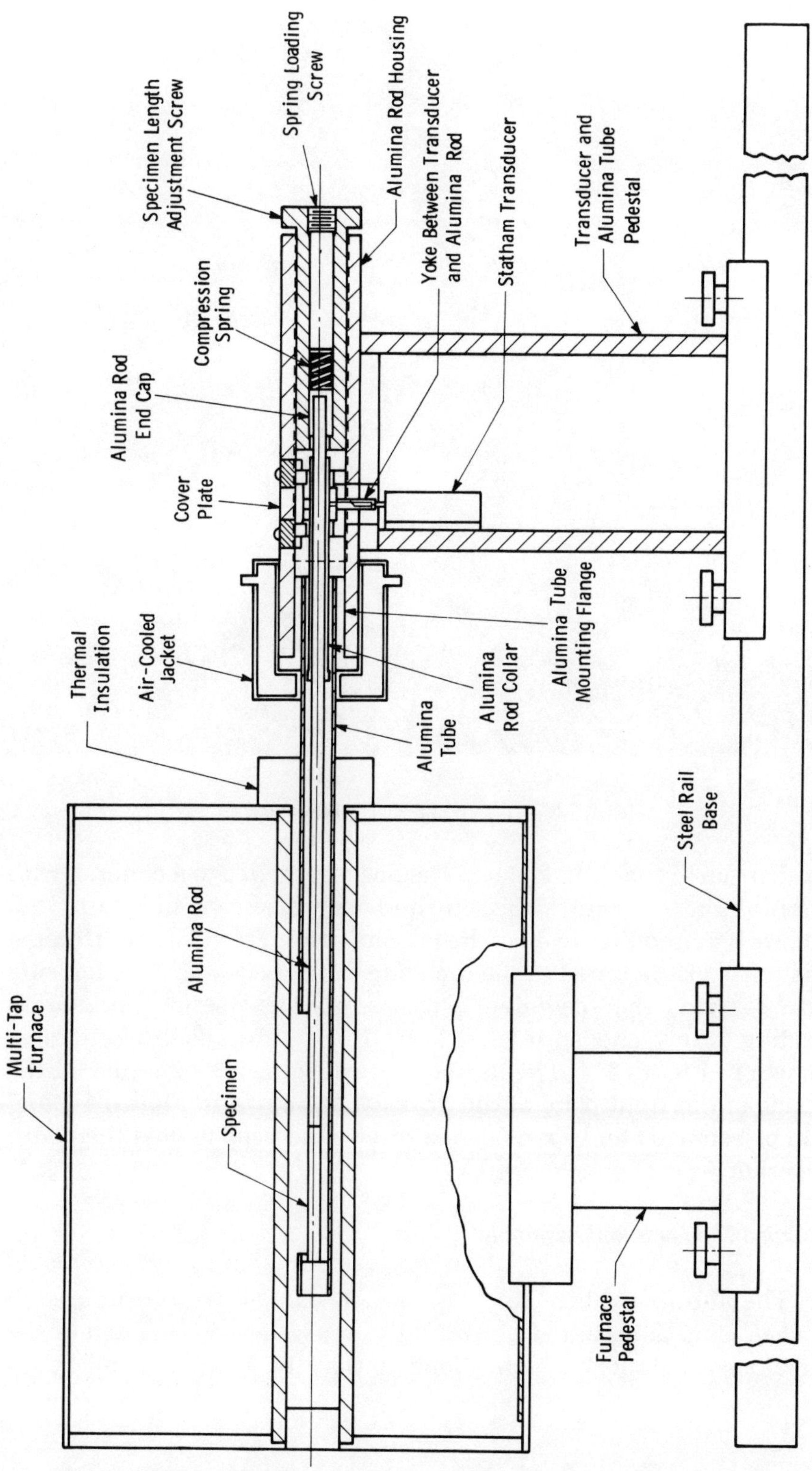

FIGURE 4–1.—Cutaway view of alumina tube dilatometer.

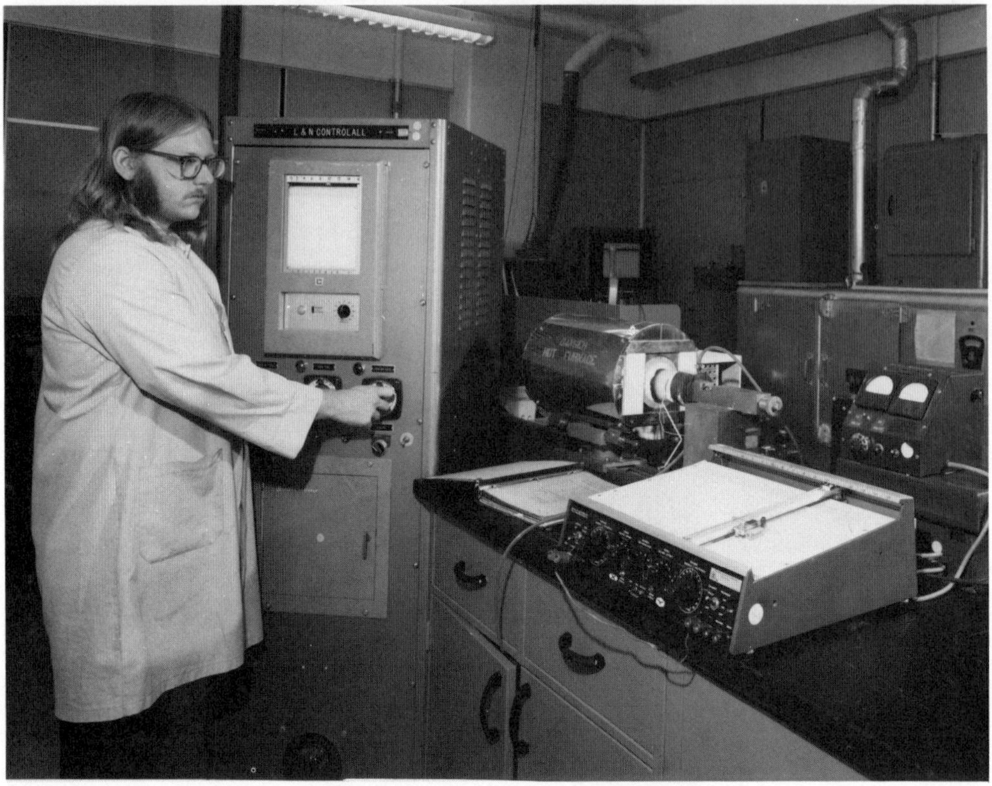

FIGURE 4–2.—Alumina tube dilatometer.

made from ⅜ inch-thick plate. The box is attached to a channel with screws, and the unit slides on the base, as shown in Figure 4-1, Figure 4-2, and Figure 4-3. The housing through which the alumina rod passes is mounted to the top plate of the pedestal. This housing also contains the specimen length adjustment screw, the spring loading screw and spring, and the Invar tube mounting flange, shown in Figure 4-4. The measuring transducer is mounted on the inside of the front plate of the pedestal as shown in Figure 4-1 and can be removed for repairs, since screws are used to hold the plates together.

Alumina Tube and Rod Assembly

The alumina tube, which is ¾ inch o.d. by ½ inch i.d. by 15 inches long, is mounted horizontally on a pedestal. One end of the tube has a cutout 3¼ inches long so that the specimen can be in-

FIGURE 4–3.—Specimen in alumina tube dilatometer.

serted; the opposite open end has an invar flange cemented to it which is used to attach the tube to the square housing above the transducer, as shown in Figure 4-1 through Figure 4-4. An invar tube $^{13}/_{32}$ inch o.d. by $4\frac{1}{2}$ inches long is cemented to the $\frac{1}{4}$ inch-diameter by $17\frac{1}{4}$ inch-long rod, as shown in Figure 4-4. In the tube the end of the alumina rod contacts the free end of the specimen; a spring load of about 3 ounces is applied to the capped end to keep the rod, specimen, and bottom of the tube in good contact at all times. A yoke from the measuring transducer is fixed to the invar tube (which is cemented to the rod) with a screw as shown in Figure 4-1 and Figure 4-3. As the specimen expands, it pushes the rod; the rod movement is transmitted to the measuring transducer through the yoke. An air-cooled jacket is mounted around the tube at the transducer pedestal to keep the transducer at room temperature. Originally it was planned to use water as the cooling medium but later it was found that air cooling was more economical and simpler. The

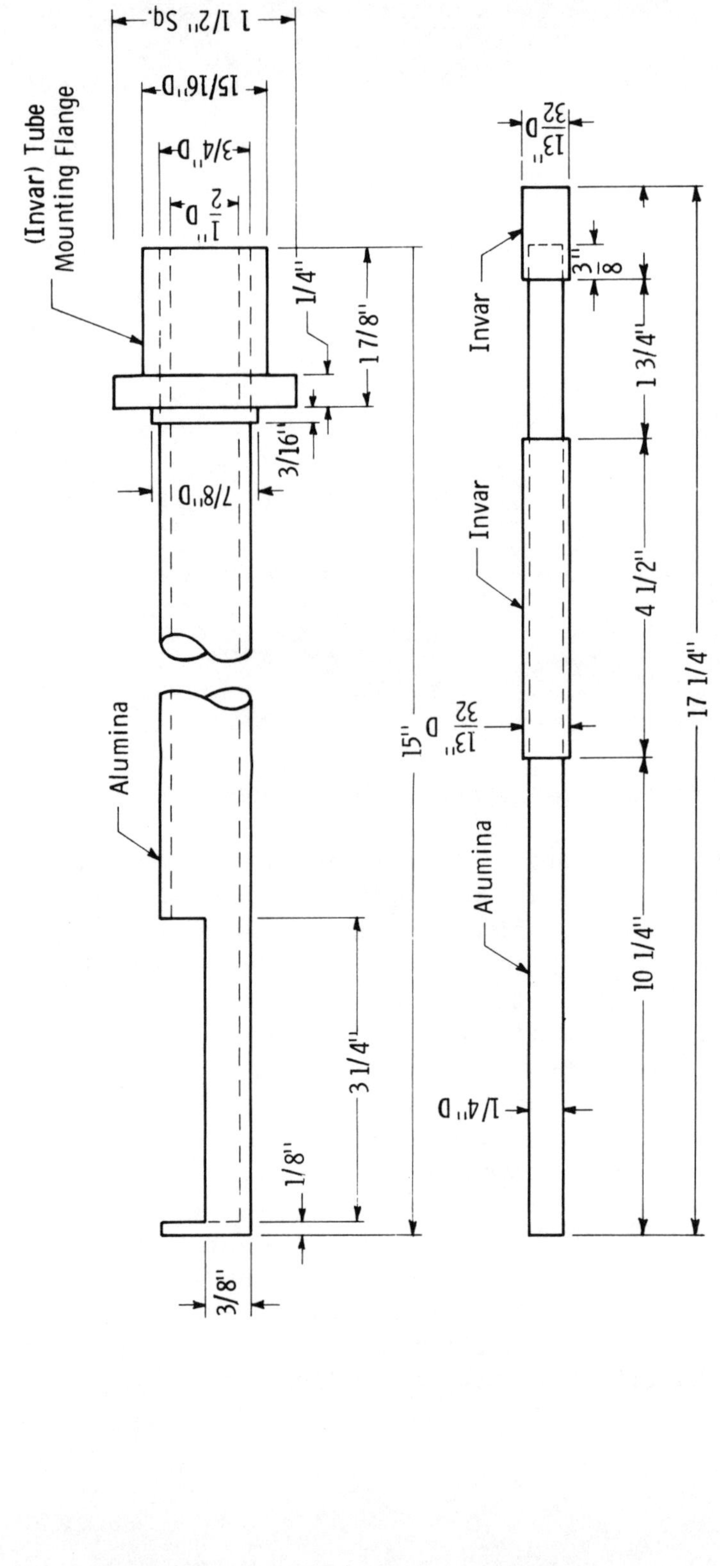

Note: All Invar Parts Cemented to Alumina Rod with Micro-Measurements, Inc. 610 Cement

FIGURE 4–4.—Alumina tube and rod dimensions.

cooling chamber is a 3-inch long, 3-inch o.d., brass cylinder with a 0.02-inch wall. The inside wall of the cylinder fits snugly around the tube. The cylinder fits snugly around the housing through which the alumina rod passes.

Although the alumina tube dilatometer was designed primarily for testing materials in air, provision is made to flow an inert gas, such as argon, into the tube for cases in which some protection against oxidation is desired. The transducer cover plate, shown in Figure 4-2 and Figure 4-3, is made with a pipe connection through which the inert gas can be introduced. The gas flows through the alumina rod housing down the tube to the specimen and out into the furnace.

Transducer, Recorder, and Thermocouple

The dimensions of the displacement transducer used to measure the thermal expansion are given in Figure 4-5. The yoke attached to the invar tube, which is cemented to the rod, contacts the transducer arm, transmitting the motion from the rod to the transducer.

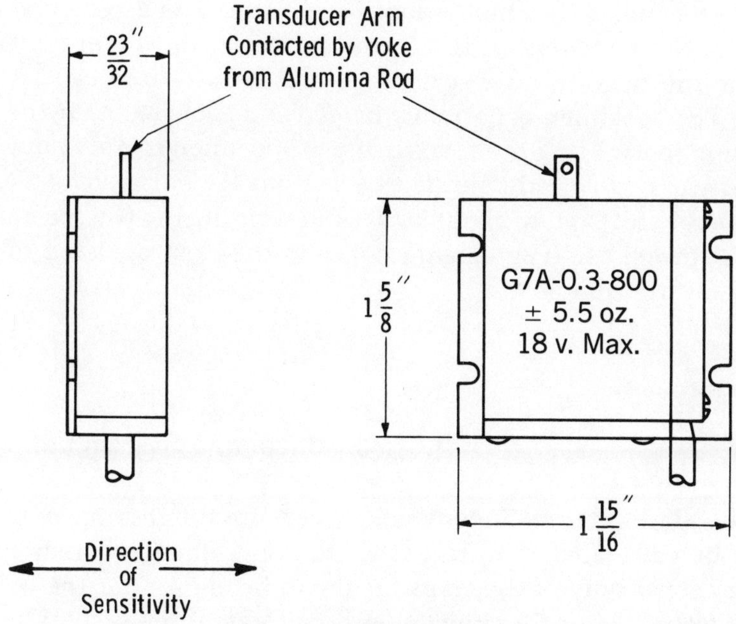

FIGURE 4–5.—Model G7A-0.3-800 Statham displacement transducer for thermal expansion measurement.

The transducer is excited by the dc power supply shown in Figure 4-2, and the excitation voltage level determines the sensitivity of the transducer. The transducer is mounted on the inside wall of the transducer pedestal, as shown in Figure 4-1, and is kept cool by the air-cooled jacket. The excitation voltage ranges from 1 to 17 Vdc in 1-V increments. The output of the transducer is recorded on the X-axis of a Hewlett-Packard Model 7001A X–Y recorder. The temperature of the specimen is measured with a platinum: platinum-10% rhodium thermocouple; its output is recorded on the Y-axis of the recorder. The resultant curve is a continuous plot of temperature vs. thermal expansion.

Furnace and Controller

The furnace is a multitap 10-zone Marshall furnace with a platinum resistance heating element. It is 10 inches in diameter by 16 inches long with a 1-inch diameter center hole for the alumina tube. The other end of the furnace is closed. During testing, the center of the specimen is located at the center of the furnace which has an 8-inch-long uniform temperature zone. The furnace temperature is controlled by a motor-controlled variac and is recorded using an L & N Speedomax "H" recorder. With the motor-controlled variac, the heating rate, soak time, and cooling rate can be regulated. The specimen is normally heated and cooled at a rate of about 5°C/min., permitting the testing of one specimen in an 8-hour shift. Because of its mass, the furnace cools from 1600°C to about 200°C in the 8-hour shift and to about 50°C sometime during the second shift and is turned off. The cooling curve is then extrapolated to room temperature from 50°C.

Transducer and System Calibration

The Statham transducer used to measure the thermal expansion must be calibrated at different excitation voltages so that the optimum sensitivity of the transducer can be chosen for the material being tested. After the transducer is calibrated, the system must be calibrated to take into account any other system errors which would result in inaccurate data. The transducer is calibrated using a micrometer or a Tuckerman gage, and the system is calibrated with a secondary platinum standard.

Transducer Calibration

As indicated in Figure 4-6, the calibration of the Statham transducer can be made using a Tuckerman optical strain gage graduated to 4×10^{-6} inches or a micrometer graduated to 0.0001 inch. It is generally easier to use the micrometer, but the Tuckerman gage is more sensitive. The special mounting head shown in Figure 4-6 is used in conjunction with these gages.

For transducer calibration, a compression spring and dummy specimen are placed in the tube as shown in Figure 4-6(B). The spring is compressed by 25% more than the expansion for which it is desired to calibrate, using the screw on the end of the mounting head if the Tuckerman gage is used or by advancing the micrometer in the direction of compression if the micrometer is used. The screw or micrometer replaces the spring loading screw of the basic dilatometer and pushes on the central plunger of the head which contacts the alumina rod and hence the compression spring through the dummy specimen. The Tuckerman gage must be read with an autocollimator of the type shown in Figure 4-6(A). The micrometer can be read directly.

During calibration the output of the transducer is fed into the *X–Y* recorder, with the excitation voltage to the transducer set at a predetermined value. The gage, Tuckerman or micrometer, is then moved in increments in the direction opposite compression (simulating thermal expansion), and the deflection is noted on the *X–Y* recorder. After the desired displacement range is covered by traversing the recorder in the expansion direction, the gage movement is reversed (simulating specimen contraction) and the incremental readings are made. With this procedure thermal expansion and thermal contraction are simulated at a fixed transducer excitation voltage and *X–Y* recorder setting. This same procedure is used for two *X–Y* recorder sensitivities at 1-V excitation voltage increments from 1 to 17V. The resulting calibration curves are shown in Figure 4-7 through Figure 4-11. Using these curves, one can select the optimum thermal expansion range for a specimen. The Statham Transducer has a range of 0.032 inch, which is sufficient for most materials. If additional range is needed, the transducer can be reset during a test.

A. Tuckerman Gage Calibration System

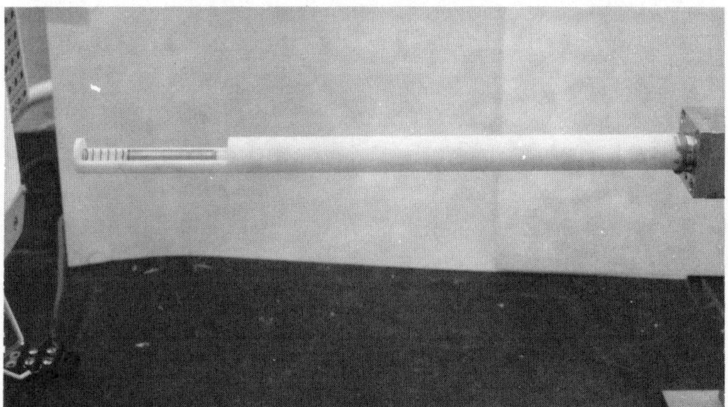

B. Compression Spring with Dummy Specimen

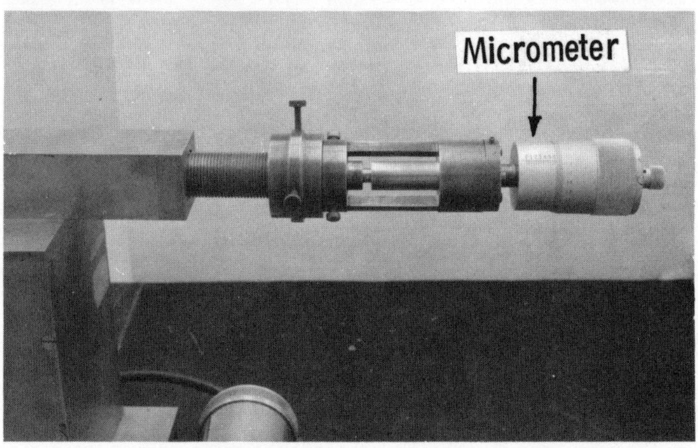

C. Micrometer Calibration System

FIGURE 4–6.—Statham transducer calibration with Tuckerman gage or micrometer.

The slopes of the curves in Figure 4-7 through Figure 4-11 are computed and summarized in Table 4-1 for easy daily reference in choosing the optimum thermal expansion range for a specimen. The data in Table 4-1 and Figure 4-7 through Figure 4-11 are based on using millimeter graph paper on the X–Y recorder.

System Calibration

After the Statham transducer is calibrated, the entire system is calibrated from room temperature to 1600°C using a chemically pure platinum standard whose thermal expansion is given in references 3 and 4. Chemically pure platinum is the standard recommended by ASTM. The specimen is heated and cooled at an average rate of

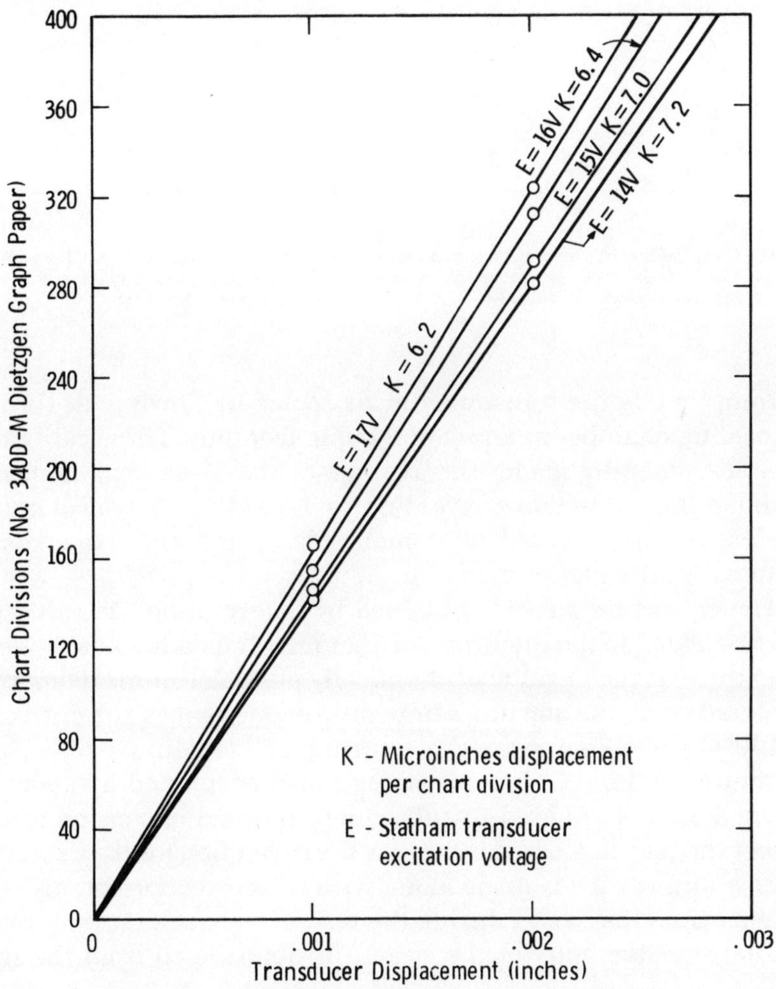

FIGURE 4–7.—Calibration of Statham transducer no. 933 to 0.003-in. displacement on X-Y recorder 15 mV range.

Table 4–1
Calibration Summary from Figure 4–7 through Figure 4–11 for a Statham Transducer Used with an Alumina Tube Dilatometer

X–Y RECORDER 15 mV FULL SCALE		*X–Y* RECORDER 75 mV FULL SCALE	
Transducer Excitation (Vdc)	Micro Inches per Chart Division	Transducer Excitation (Vdc)	Micro Inches per Chart Division
1	105.9	1	—
2	53.0	2	—
3	35.7	3	—
4	26.8	4	132.5
5	21.1	5	108.7
6	15.5	6	88.7
7	15.0	7	78.7
8	12.5	8	65.5
9	11.5	9	58.1
10	10.2	10	53.5
11	9.3	11	47.5
12	8.7	12	42.9
13	7.7	13	40.0
14	7.2	14	36.8
15	7.0	15	35.0
16	6.4	16	32.4
17	6.2	17	—

NOTE: These calibrations were made with a Hewlett-Packard Moseley Autograph *X–Y* Recorder Model 7001A using Dietzgen graph paper no. 340D-M, dc power supply no. 2, and a Statham Transducer Model No. G7A-0.3-800 (Transducer Serial No. 933).

5°C/min., while the transducer is air cooled by flowing air through the cooling chamber at a rate of 2 cubic feet/min. Three calibration tests are generally made; the average of the three tests is used to establish the correction curve. Figure 4-12 shows a typical correction curve, uncorrected dilatometer data, and the expansion of chemically pure platinum.

The correction curve is obtained by determining the value that must be added to the uncorrected dilatometer data to make it identical to the expansion value of chemically pure platinum. Using room temperature as the starting temperature, one makes the correction at 100°C increments on the heating and cooling cycles, to a maximum of 1600°C. These values are then plotted as shown in Figure 4-12 and are used in subsequent thermal expansion tests to correct the test data. By this process the correction for the expansion of the alumina tube is made along with the correction for any other system errors that occur during the test.

The reproducibility of the system is determined from the three

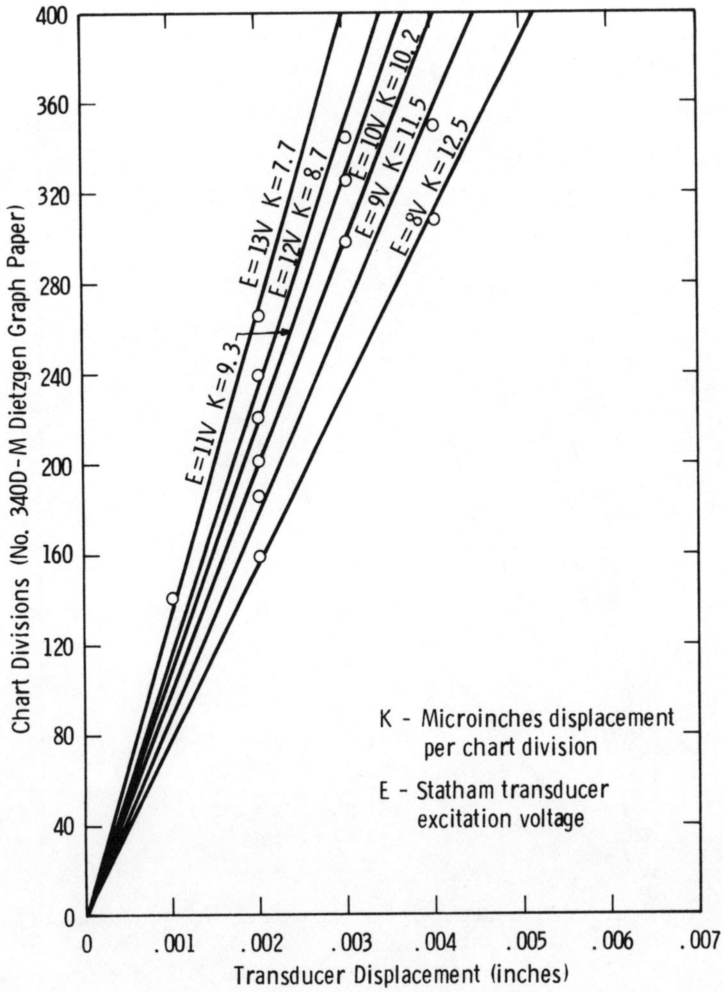

FIGURE 4–8.—Calibration of Statham transducer no. 933 to 0.007-in. displacement on X-Y recorder 15 mV range.

calibration tests. As might be expected, some hysteresis occurs, as is shown in Figure 4-12; so the correction for cooling is slightly different from that for heating. This is of no consequence as long as the test is repeatable. Similar but smaller differences are found in vitreous silica tube and rod systems. Alumina has about 18 times the expansion of vitreous silica and hence is more sensitive to thermal differences during heating and cooling resulting in larger correction values. This is also true of the tantalum tube dilatometer described in Chapter 5 and reference 2 except that the factor is 13 rather than 18.

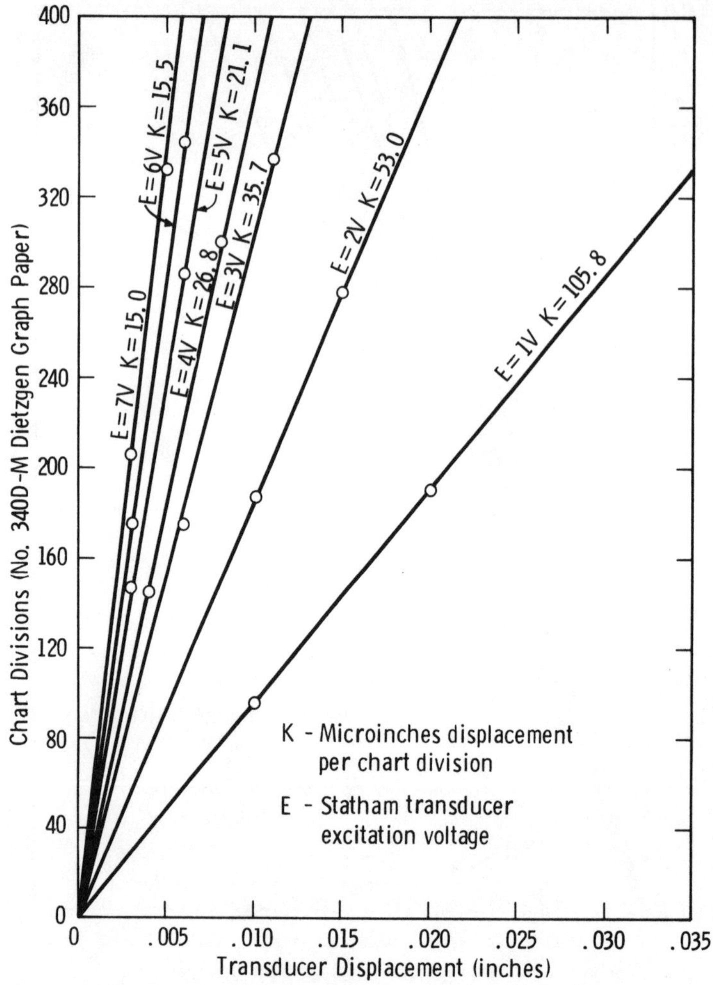

FIGURE 4–9.—Calibration of Statham transducer no. 933 to 0.035-in. displacement on X-Y recorder 15 mV range.

Testing Procedure

After the Statham displacement transducer and the dilatometer are calibrated, specimens can be tested and the data corrected. However, a daily check of the *X–Y* recorder must be made before every test because its calibration can change. The *X–Y* recorder is electrically calibrated on both axes before every test using an L & N No. 8662 precision potentiometer or equivalent. This is accomplished by feeding millivolt signals from the potentiometer into each

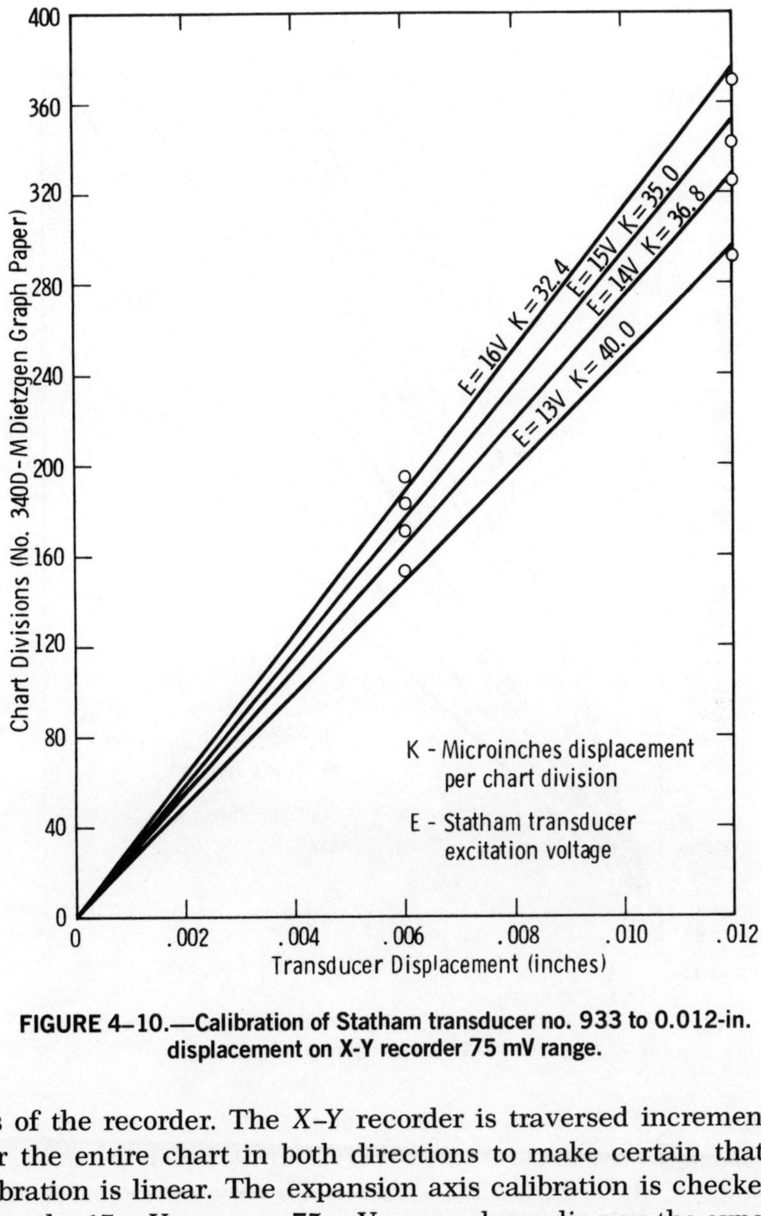

FIGURE 4–10.—Calibration of Statham transducer no. 933 to 0.012-in. displacement on X-Y recorder 75 mV range.

axis of the recorder. The *X–Y* recorder is traversed incrementally over the entire chart in both directions to make certain that the calibration is linear. The expansion axis calibration is checked on either the 15-mV range or 75-mV range, depending on the expected expansion; the temperature axis is calibrated in the 20-mV range to obtain the maximum deflection across the recorder for the 25°C (room temperature) to 1600°C temperature range, which represents a change of 16.573 mV. During the test the temperature axis is checked with the potentiometer to make certain that any minor erra-

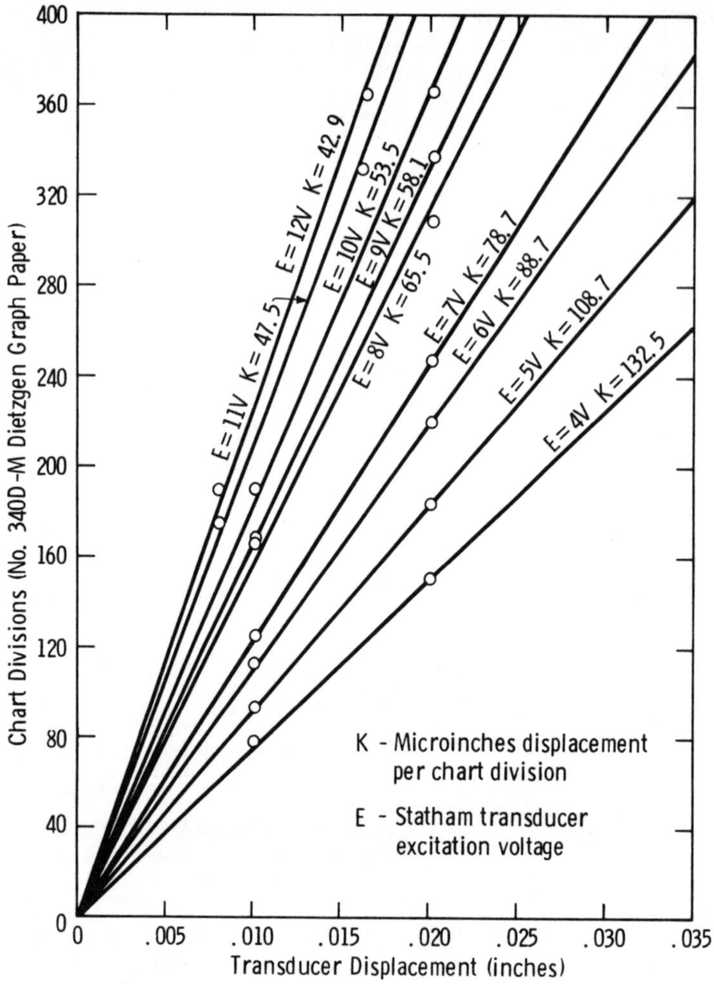

FIGURE 4–11.—Calibration of Statham transducer no. 933 to 0.035-in. displacement on X-Y recorder 75 mV range.

tic behavior of the X–Y recorder is identified and corrected. In general, the X–Y recorders function well; if erratic behavior occurs, it is generally obvious from a discontinuity in the curve of thermal expansion vs. temperature.

Results

The accuracy of the dilatometer was determined by measuring the thermal expansion of alumina, Lucalox, magnesia, and thoria

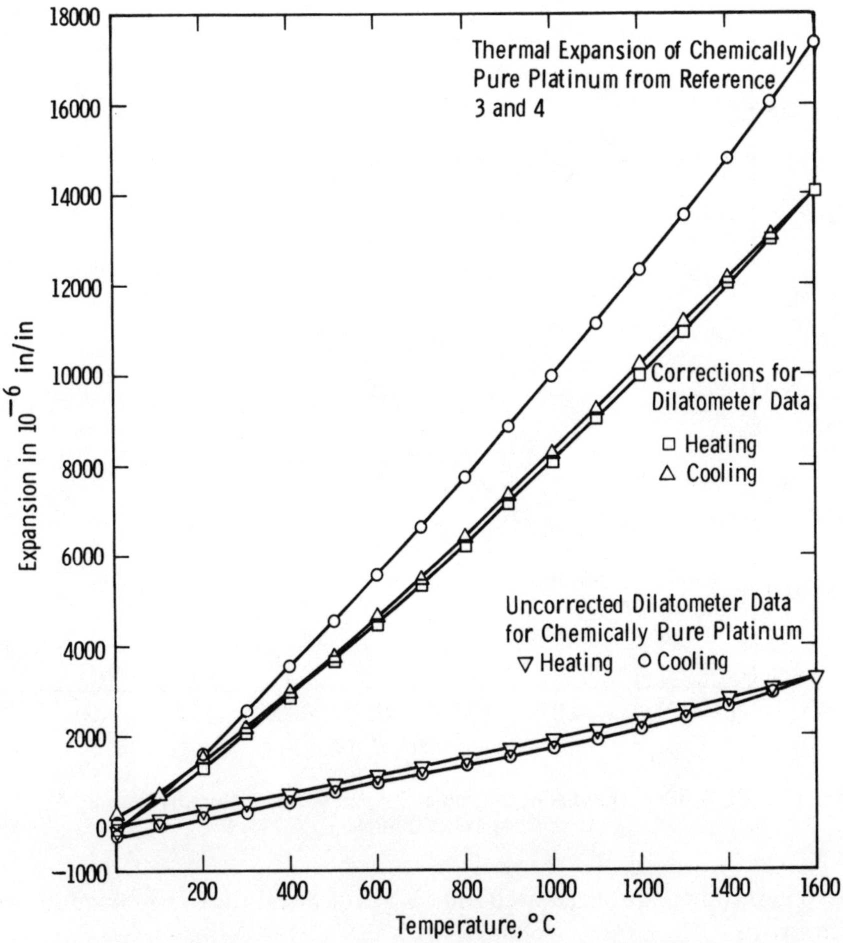

FIGURE 4–12.—Thermal expansion of chemically pure platinum, uncorrected dilatometer data for chemically pure platinum, and corrections for the dilatometer data.

from room temperature to 1600°C in air; the test results were compared to literature values. The data for these materials are presented graphically in Figures 4-13 through Figure 4-16 for both the heating and cooling cycles. The results obtained for alumina, magnesia, and thoria were compared to the data reported by Campbell[5], and the Lucalox results were compared to the data reported by Matheson[6].

As measured in this survey, the thermal expansion of alumina and thoria during the heating and cooling cycles, shown in Figure 4-13 and Figure 4-14, duplicates the results reported by Campbell[5]. Any minor variations are probably the result of impurity differences

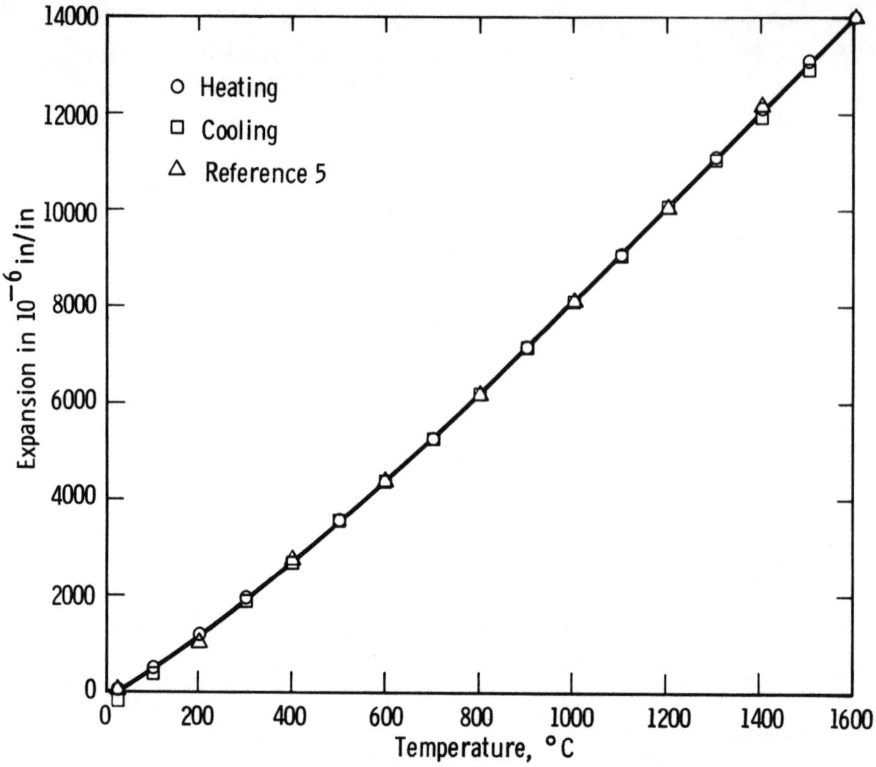

FIGURE 4–13.—Thermal expansion of 99 percent pure alumina from 25°C to 1600°C in air.

between the specimen tested in this survey and those reported in the literature.

The heating and cooling cycles for magnesia, shown in Figure 4-15, are not exactly the same. This is also true for Lucalox, as shown in Figure 4-16. The results obtained for magnesia during the heating cycle in this survey are the same as those reported in the literature which were also probably measured during the heating cycle. The results of the heating cycle are slightly higher than those of the cooling cycle. The specimen tested in this survey exhibited a slight permanent contraction on the cooling cycle, which could be the result of additional thermal instability. For Lucalox, the results of the heating and cooling cycles are closer than those for magnesia. The results reported in the literature for Lucalox are about the same as those found in this survey to 1000°C. Comparative results above 1000°C are not available. The continuation of the same general slope

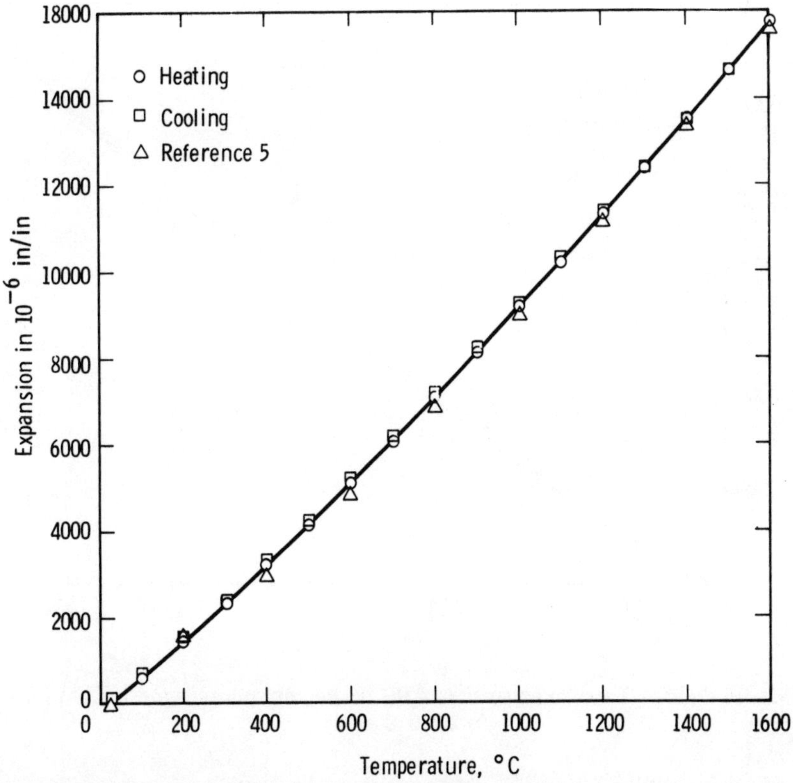

FIGURE 4–14.—Thermal expansion of 98.5 percent pure thoria from 25°C to 1600°C in air.

for the data obtained in this survey from room temperature to 1000°C to the higher temperature indicates that the data obtained in this survey are highly accurate.

Discussion

The thermal expansion results reported here were for both the heating and cooling cycles. Most data reported in the literature are for the heating cycle only, because it is difficult to program a dilatometer to make the heating cycle data coincide with the cooling cycle data. Each dilatometer must be treated individually to establish the proper procedure, and techniques must be developed using a

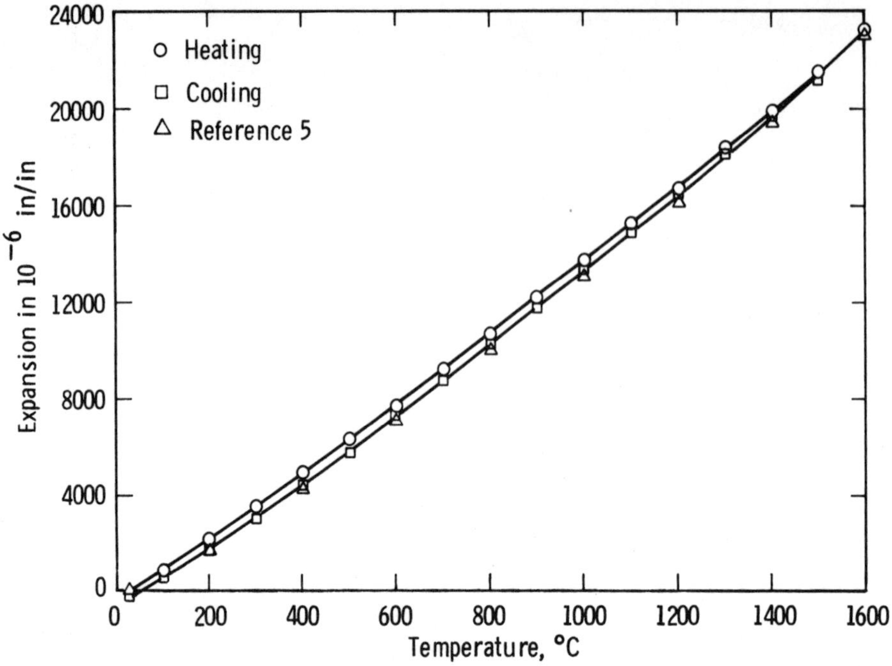

FIGURE 4–15.—Thermal expansion of 99.99 percent pure magnesia from 25°C to 1600°C in air.

standard material such as platinum, whose expansion and contraction are identical. The technician can develop the proper techniques to make the heating cycle data coincide with the cooling cycle data by running several tests with the platinum standard and then applying the same procedure for test specimens. In general, hysteresis occurs at the top temperature, when the controller is turned around from heating to cooling. The hysteresis can be eliminated by cutting back the power to the furnace as the top temperature is approached.

No effort was made to place a percentage figure on the accuracy of the dilatometer results reported here because greater specimen-to-specimen variation in the thermal expansion of oxides can occur than indicated in this survey. Incomplete sintering, impurity variations, and density variations can cause thermal expansion variations that may account for the small differences cited in this survey. During the calibration of the dilatometer, using platinum as the standard, the author obtained accuracies on the order of ±3%.

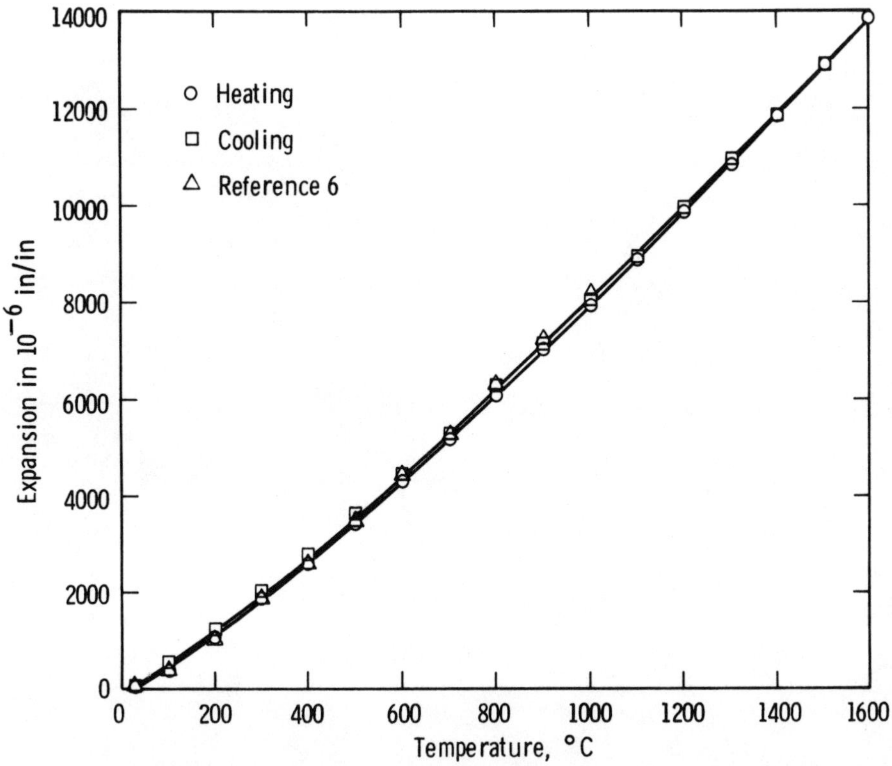

FIGURE 4–16.—Thermal expansion of Lucalox from 25°C to 1600°C in air.

References

1. J. Valentich, "New Values for Thermal Coefficients." *Product Engineering*, July 19, 1965.
2. J. Valentich, "Thermal Expansion Measurement." *Instrument and Control Systems*, vol. 42, Oct. 1969, pp. 91-94.
3. ASTM. Designation E228 *Linear Thermal Expansion of Rigid Solids with a Vitreous Silica Dilatometer.*
4. J. L. White, "The Suitability of Platinum, Molybdenum, Tantalum, and Tungsten as High-Temperature Length Standards." *NRL Report 5159, Washington, D.C. Naval Research Laboratory, June 17, 1958.*
5. I. E. Campbell, *High-Temperature Technology*, John Wiley & Sons, New York, 1956.
6. R. R. Matheson, "G.E.'s Lamp Research Leads to Unique Ceramic Material." *Ceramic Age*, June, 1963.

CHAPTER 5

Dilatometer for Measurements from 25°C to 2200°C in Vacuum

Introduction

The thermal expansion of most engineering materials is determined from cryogenic temperatures to about 1000°C in an air atmosphere using the vitreous silica tube dilatometer described in Chapter 3. The useful upper temperature for ceramic materials is about 1600°C; and the alumina tube dilatometer described in Chapter 4 was designed to make thermal expansion measurements to 1600°C in an air atmosphere. Refractory materials and space oriented materials are subject to temperatures above 1600°C in a vacuum or inert gas atmosphere. To measure the thermal expansion of these materials the tantalum tube dilatometer described in this chapter was designed and built for measurements to 2200°C.

Many types of dilatometers were investigated in the literature to determine which would best serve the needs of testing to about 2200°C. The principal items of interest in evaluating the various dilatometers were good accuracy, low initial cost, minimum of auxiliary equipment, low maintenance, high service temperature, and variable specimen size. No one dilatometer could be found which fulfilled all these requirements, but one, made by the Argonne National Laboratory, was found which fulfilled many of them. The dilatometer described in this chapter is similar to the Argonne apparatus except that the specimen holder was completely redesigned and some changes were made to accommodate existing auxiliary equipment.

Most dilatometers in use today use inert gas under positive pressure instead of a vacuum atmosphere because fewer problems are encountered. This is especially true when testing graphite, which deposits out on the internal components. The apparatus described here, however, is more universal than others because either vacuum or inert gas can be used.

Theory

The mean coefficient of linear thermal expansion α_m between temperatures T_t and T_0 is defined by the following equation:

$$\alpha_m = \frac{L_t - L_o}{L_o (T_t - T_o)} = \frac{1}{L_o} \frac{\Delta L}{\Delta T} \tag{1}$$

where

α_m = the mean coefficient in inches/inch degree

L_t and L_o = specimen length at temperature T_t and T_o respectively, in inches

and

$$\Delta L = L_t - L_o$$

$$\Delta T = T_t - T_o$$

For convenience let the reference temperature T_o be 0°C; then equation (1) can be written as

$$\alpha_m = \frac{1}{L_o} \frac{\Delta L}{T} \tag{2}$$

The length of the specimen is then

$$L = L_o\,(1 + \alpha_m T) \tag{3}$$

The instantaneous coefficient of linear thermal expansion α_1 is defined by the following equation:

$$\alpha_i = \underset{T_t \to T_o}{\text{Limit}} \frac{L_t - L_0}{L(T_t - T_o)} = \frac{1}{L}\frac{dL}{dT} \tag{4}$$

where $\frac{dL}{dT}$ = slope of the length-temperature curve in inches/inch degree

From equation (3)

$$\frac{dL}{dT} = \left[\alpha_m + T\left(\frac{d\alpha_m}{dT}\right)\right] L_o \tag{5}$$

Substituting (3) and (5) into (4)

$$\alpha_i = \frac{\alpha_m + T\dfrac{d\alpha_m}{dT}}{1 + \alpha_m T} \tag{6}$$

which is the relationship between α_i and α_m

It is common practice to present thermal expansion data in the form of a continuous curve with the thermal expansion in inches/inch along the ordinate, and the temperature in °C along the abscissa. The mean or instantaneous coefficients can then be calculated from the above expressions. The instantaneous coefficients will generally be higher than the average coefficients since the thermal expansion curves are usually concave upward.

Dilatometer Components

The high temperature dilatometer consists of a gas-tight, water-cooled chamber containing a tantalum resistance heating element, radiation shields, push rod and tube. A linear variable differential transformer (LVDT) attached to the tantalum push rod which contacts the specimen is used to measure the thermal expansion. The expansion is recorded as a function of temperature on an *X–Y* recorder.

Chamber Construction

Figure 5-1 is an assembly view of the dilatometer and Figure 5-2 is a schematic view of the dilatometer and its auxiliary equipment. The thermal expansion system with a specimen under test is shown in Figure 5-3. The sides, bottom, and top of the dilatometer are of double wall construction and are water cooled. The inside walls are made from stainless steel, and the outside walls from hot rolled steel. The copper leads supplying current to the tantalum electrodes are also water cooled. The heating element is a tantalum sleeve resistance heater. Heat is confined to the specimen by a series of four concentric, cylindrical tantalum radiation shields and four tantalum cups at the top and the bottom of the concentric shields. Appropriate holes are cut in the shields for the temperature control thermocouple and for the insertion of the tantalum tube, which contains the specimen.

Measurement of Specimen Elongation

Movement of the tantalum push rod, which is a measure of specimen elongation, is sensed by means of a Daytronic Model 102-A-120 linear variable differential transformer. The output from the transformer is fed into a Daytronic Model 300BF differential transformer indicator and amplifier and then into one axis of a Hewlett-Packard Model 7001A autographic recorder. The LVDT is anchored to an adjustable platform made from invar. An adjusting screw, also made from invar, raises or lowers the platform to accommodate different sized specimens. Invar with its low coefficient of thermal expansion reduces to a minimum any extraneous expansion that would cause error in the thermal expansion measurements. Efficient water cooling keeps the temperature rise of the invar platform to a maximum of 5°C above room temperature when the specimen is at 2200°C; this slight temperature rise introduces negligible error in the expansion measurements.

Tantalum Tube

The tantalum tube shown in Figure 5-4 holds the specimen. It has a ⅝-inch o.d., a 0.02-inch wall and is 10 inches long. Holes ¼

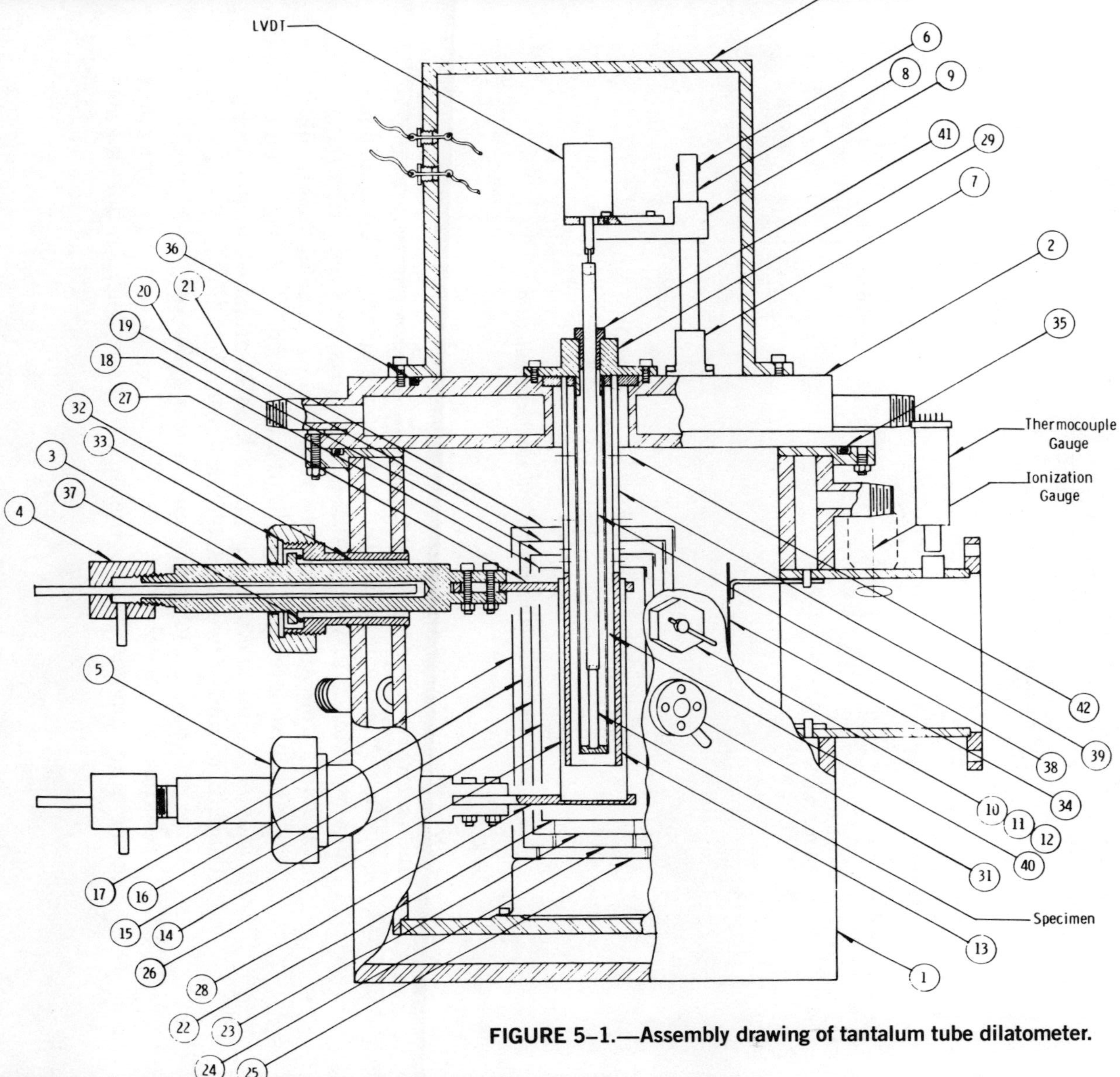

FIGURE 5–1.—Assembly drawing of tantalum tube dilatometer.

ITEM	MAT'L	NAME
1	Steel & Stn. Stl.	Body
2	Steel & Stn. Stl.	Head
3	Copper	Electrode
4	Brass	Water Supply
5	Brass	Cap Nut
6	Invar	Adj. Screw
7	Invar	Trans. Support
8	Steel	Guides
9	Invar	Adjust. Slide
10	Stn. Stl.	Shield Shaft
11	Brass	Cap Nut
12	Stn. Stl.	Sight Shield
13	Moly	Suseptor
14-17	Tantalum	Heater Shield
18-21	Tantalum	Top Shields
22-25	Tantalum	Bottom Shields
26	Tantalum	Heater
27	Tantalum	Top Electrode
28	Tantalum	Bottom Electrode
29	Stn. Stl.	Flange
30	Steel	Transducer Cover
31	Quartz	Sight Glass
32	Teflon	Bushing
33	Teflon	Cap
34	Stn. Stl.	Vac. Shield
35-37	Rubber	"O" Rings
38	Tantalum	Push Rod
39	Tantalum	Support Rod
40	Tantalum	Tube
41	Stn. Stl.	Bushing
42	Tantalum	Tube Shields

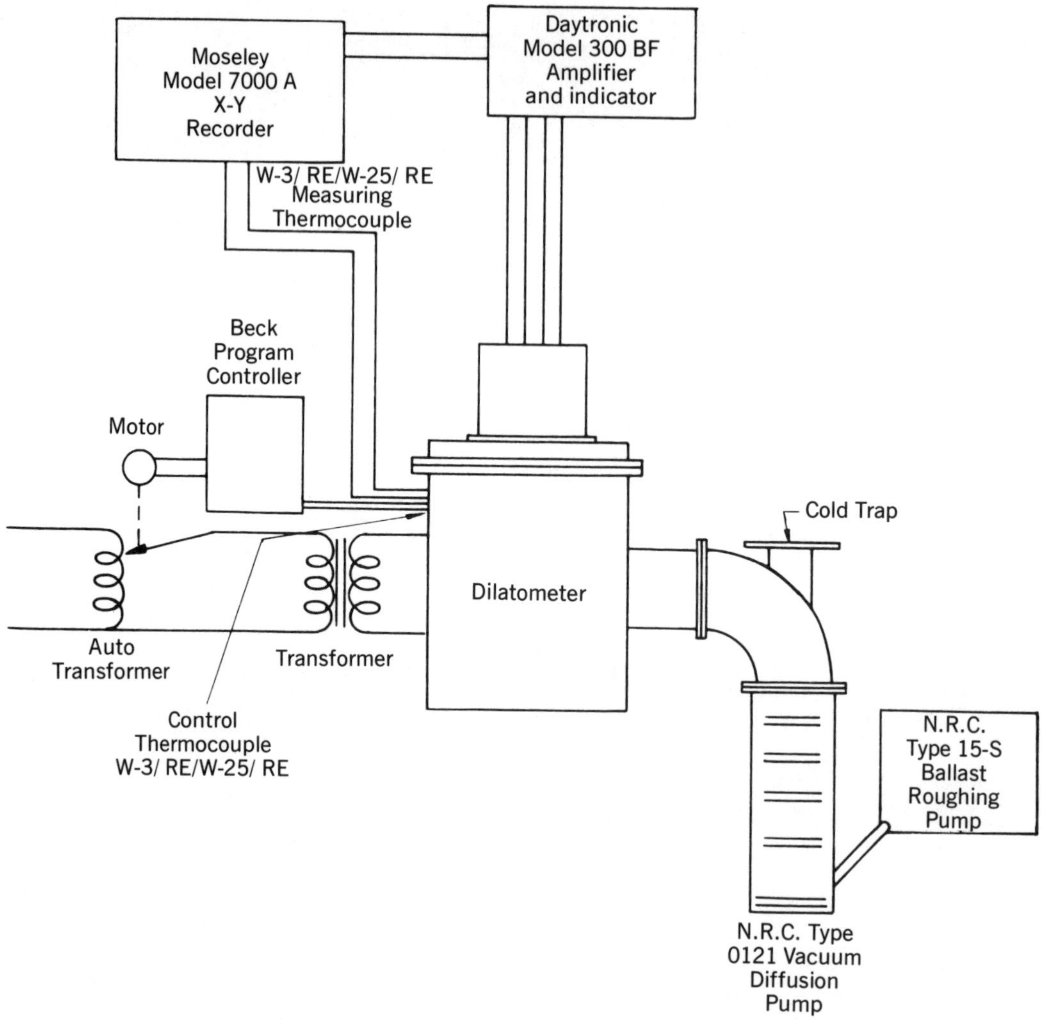

FIGURE 5–2.—Dilatometer and auxillary equipment.

inch in diameter are drilled through the tantalum tube in the area of the specimen to permit better heat flow from the heater to the specimen. A flange is press-fitted onto the open end of the tube. This flange is attached to a larger flange by screws, and the larger flange is insulated from and fixed to the water cooled head of the chamber by screws during testing, as indicated in Figure 5-1. A molybdenum susceptor 1 inch o.d. and 5 inches long is attached to the small flange by means of two ⅛-inch-diameter tantalum rods. Set screws secure the rods to the flange, and a screw connection is used between the susceptor and the rods. The susceptor serves to minimize temperature gradients along the tube and specimen. A bushing in

FIGURE 5–3.—Tantalum tube dilatometer with specimen under test.

the open end of the tube serves to guide the tantalum rod resting on top of the specimen, which is at the bottom of the tube. The 1/16-inch o.d., 2-hole thoria thermocouple insulator also passes through a hole in the bushing.

Temperature Measurement

Specimen temperature is measured with a high-temperature tungsten-3% rhenium tungsten-25% rhenium thermocouple fixed to the specimen with a tantalum band as shown in Figure 5-4. This type of thermocouple can be used to about 2300°C. Thoria insulators, which have a melting point of about 3300°C, are used to insulate the thermocouple. A W-3% Re/W-25% Re thermocouple welded to the tantalum heater and connected to a program controller is used to control the temperature change of the specimen. Both the specimen and control thermocouples are made from 0.011-inch-

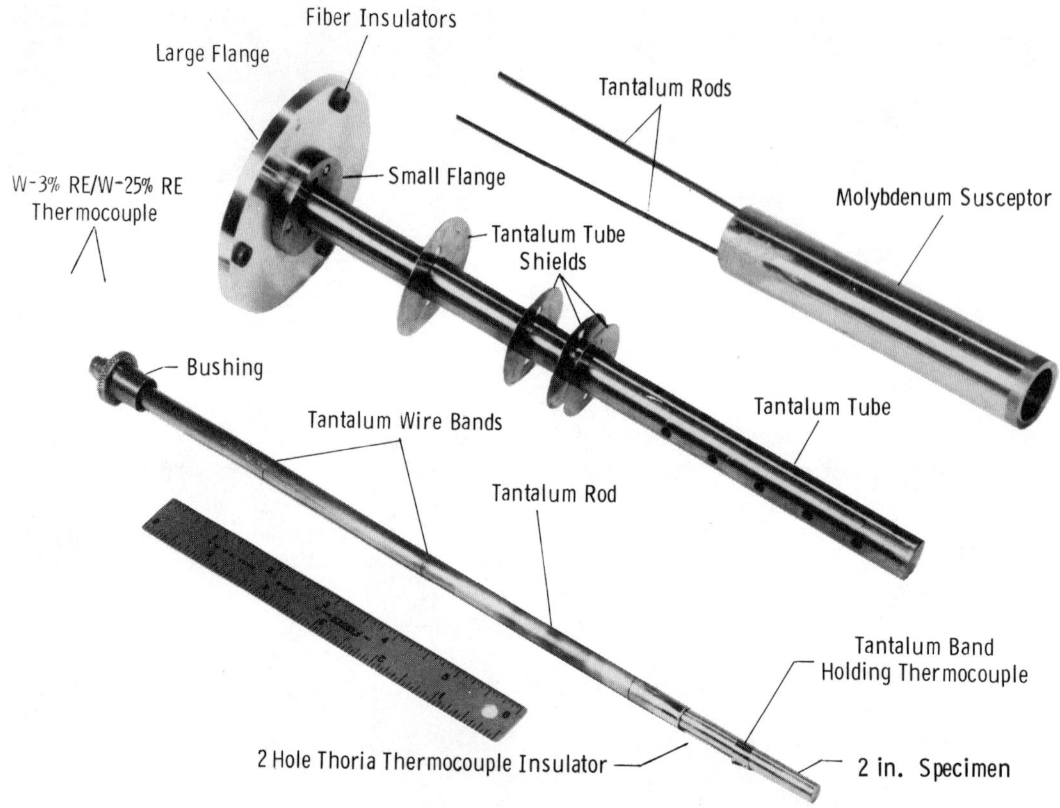

FIGURE 5–4.—Tantalum tube and rod, specimen with thermocouple and molybdenum susceptor

diameter wire, and the bead is made by heliarc welding in an argon atmosphere.

In making the bead, copper chill blocks are used to clamp the wires of the thermocouple, and an oval shaped bead is formed against the copper by the heliarc weld. The copper blocks act as a heat sink to keep the heat-affected embrittled zone as small as possible.

The temperature of the specimen is recorded on the Y-axis of the autographic recorder, and a precision Leeds and Northrup potentiometer is used to check the recorder. The Model 7001 A recorder is used because it contains a guard circuit on both axes which prevents stray electrical pick-up from affecting the signal.

Vacuum Equipment

Vacuum on the order of 1×10^{-6} torr is attained during initial start-up of the apparatus. At the high temperatures this vacuum is reduced to about 5×10^{-6} torr.

The vacuum pumping system consists of an N.R.C. Type 15-S portable roughing pump and a 4-inch N.R.C. Type 0121 portable vacuum diffusion pump with a liquid nitrogen cold trap. The piping and valving are arranged so that the chamber can be roughed down while the diffusion pump is being brought up to temperature.

Vacuum is measured by means of thermocouple gages and an ionization gage.

Power Supply

The power supply consists of a reversible motor-driven auto-transformer and a conventional transformer. The temperature rise of the specimen is controlled by a Beck program controller and is normally set at a rate of 7°C/min., but it can be increased to 35°C/min. The control thermocouple is welded to the tantalum heater. A Putmo pulse-time modulation controller, on which the rate of approach is set, operates the motor of the auto-transformer in either direction, depending upon whether the temperature of the specimen is above or below the control point. With this arrangement it is possible to measure expansion during both heating and cooling. Since the system has a low heat capacity, the cooling curve can be controlled precisely without the need of an additional external cooling medium such as an inert gas. To reach a specimen temperature of 2200°C about 7 KW of power is needed, as shown in Figure 5-5.

Calibration

In Figure 5-1 it can be seen that the tantalum push rod is longer than the tantalum tube holding the specimen. Furthermore, the tube is attached to the water-cooled head. This additional cooling and the difference in length cause a differential expansion between the two parts; a correction must be made for this expansion, otherwise the LVDT senses this movement as expansion of the specimen.

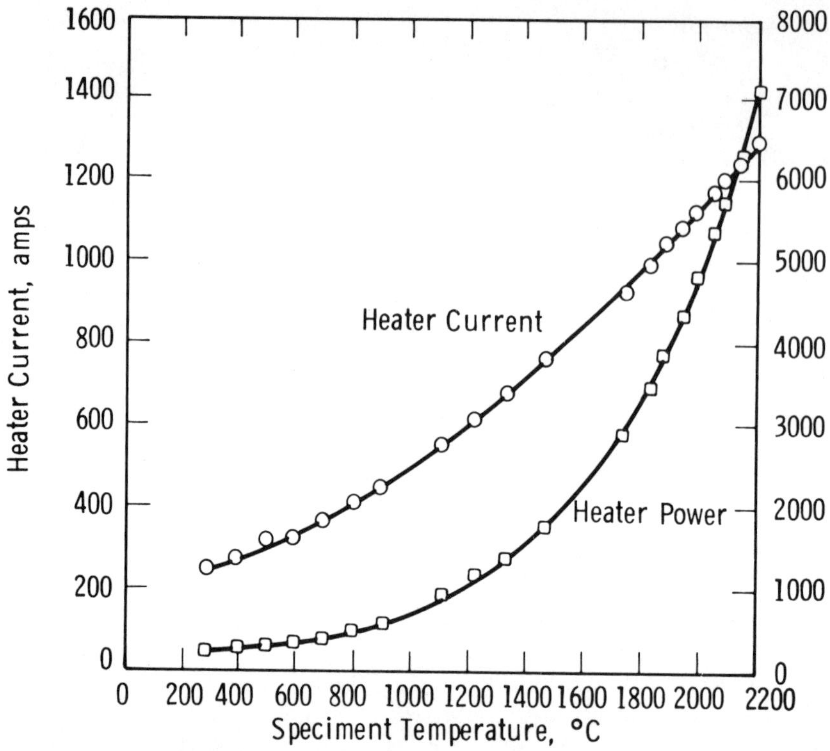

FIGURE 5–5.—Typical curves of heater current and power vs. specimen temperature.

In the section of the tantalum tube where the specimen is held, there is no tantalum rod to compensate for the expansion of the tube; therefore a correction must be made. To obtain this correction all that is necessary is to multiply the original specimen length by the coefficient of thermal expansion of the tantalum tube.

Although it should be possible to calculate the corrections involved, from a practical standpoint it is necessary to determine them experimentally. To do this, a 2-inch-long by ¼-inch-diameter pure tantalum specimen was prepared, and an apparent thermal expansion curve was obtained at a heating rate of 7°C/min. The ideal design would indicate no expansion of the specimen. But since differential expansion occurs because of the difference in lengths of the tantalum push rod and tube, and because of the effect of the water-cooled head, an apparent expansion correction must be applied to all data obtained using this apparatus. Figure 5-6 shows the expansion of pure tantalum measured by Edwards[1], and Figure 5-7 the un-

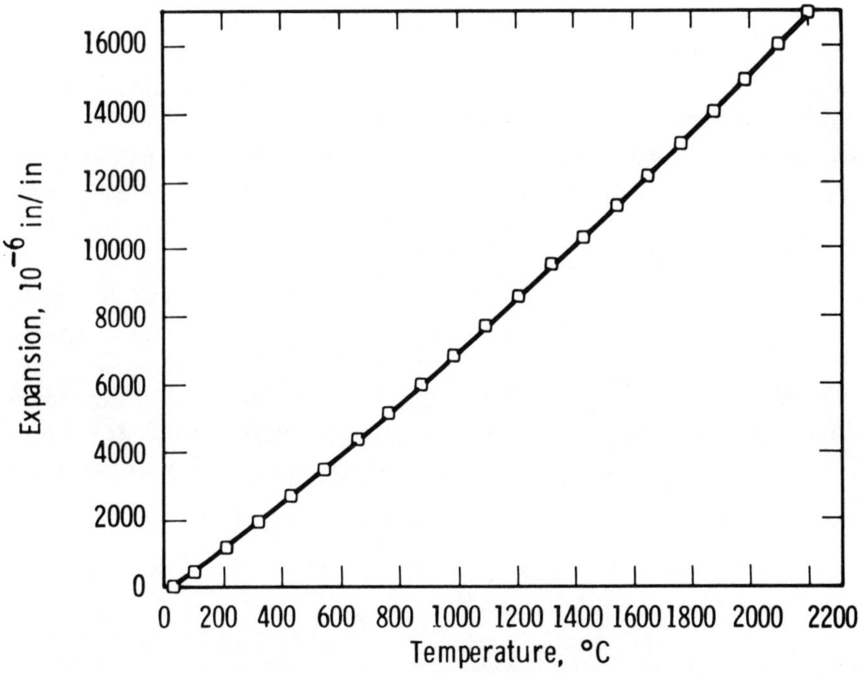

FIGURE 5–6.—Thermal expansion curve for tantalum.

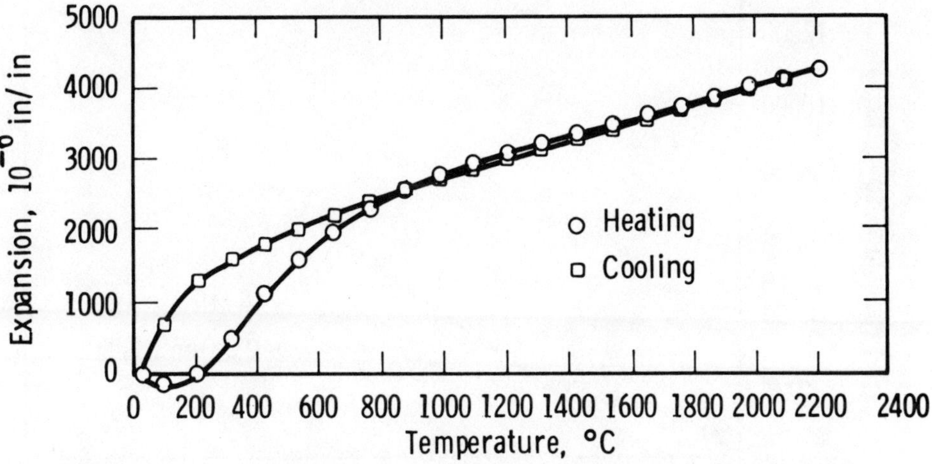

FIGURE 5–7.—Uncorrected thermal expansion curve for tantalum as measured with Westinghouse tantalum tube dilatometer.

corrected thermal expansion of tantalum measured with the dilatometer. Figure 5-8 shows the correction that must be added to the uncorrected data to obtain the true expansion of a material. The corrections are obtained by subtracting the data in Figure 5-7 from the data in Figure 5-6 at every temperature. The curve in Figure 5-8 represents two calibration tests using a pure tantalum specimen 2 inches long. The calibration data from these tests agree within 3% from 650°C to 2200°C to 650°C.

As shown in Figure 5-8, the dilatometer correction curve for the range from room temperature to 650°C is not the same for the heating and the cooling cycles. If the heating and the cooling rates were very slow in this temperature range, the curves would agree; however, it is not practical to conduct tests at such a slow rate. Therefore, expansion data for this temperature range are obtained using the vitreous silica tube dilatometer.

Test Results and Discussion

Oxides, refractory metals, and graphite have been tested in this apparatus. The data obtained for thoria to 2200°C is shown in Figure

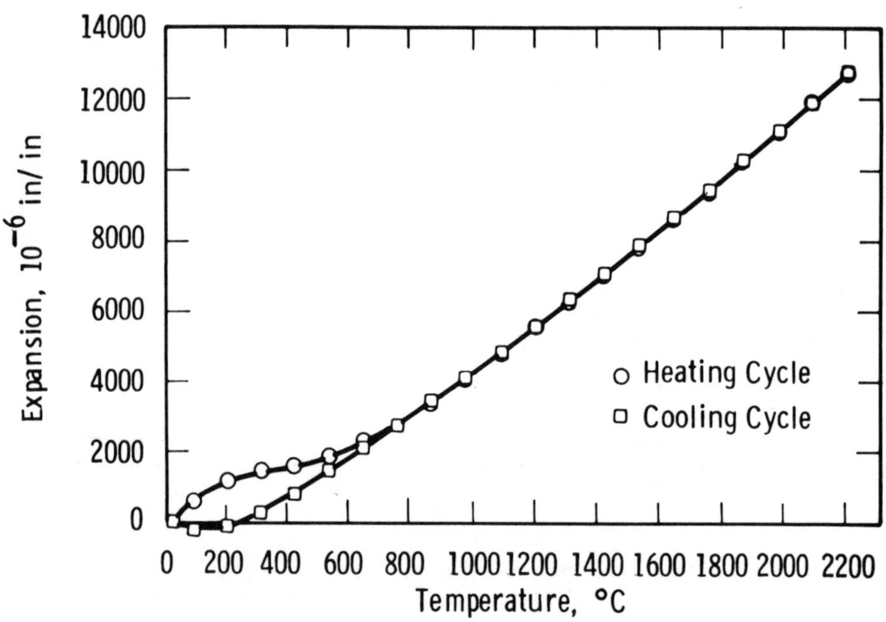

FIGURE 5–8.—Correction curve for Westinghouse tantalum tube dilatometer.

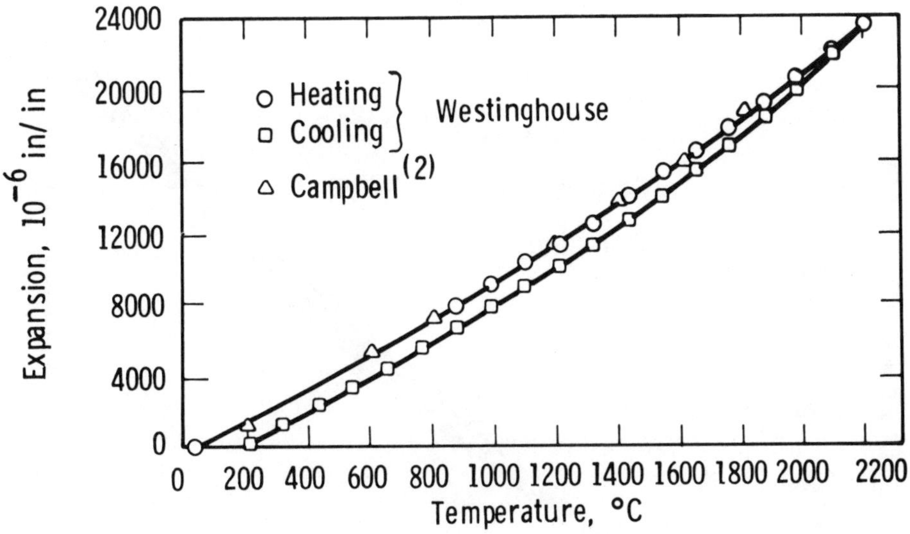

FIGURE 5–9.—Thermal expansion of thoria from 25°C to 2200°C to 25°C in vacuum.

5-9 and is within about 2% of the data reported in the literature for a heating cycle. No comparison for the data obtained on the cooling cycle could be found in the literature. The measurement of the specimen length after the test confirmed that the specimen had shrunk, as shown on the cooling curve. The shrinkage is probably due to moisture being driven out of the thoria.

The data obtained for molybdenum (99.9% pure) is shown in Figure 5-10 along with the results of Conway[3] and Worthing[4]. The data of Conway and Worthing are for the heating cycle only. Data for the cooling cycle are seldom presented in the literature because they are generally controversial and do not duplicate the heating cycle data. The data obtained in this survey are higher and more linear than those obtained by Conway and Worthing. The data of Conway have the lowest values and those of Worthing are in between. Worthing's data fall on the cooling curve obtained in this survey. The spread in all of the data of Figure 5-10 is representative of what is reported by other investigators.

The thermal expansion curves obtained for arc vacuum cast tungsten (99.9% pure) are presented in Figure 5-11. The data obtained in this survey on the heating cycle and on the cooling cycle from 2200°C to about 1100°C agree with those of Conway[3] and Worthing[5] within about 3%.

The data obtained for ZTA graphite in the cross-grain and the with-grain directions are given in Figure 5-12. The total expansion

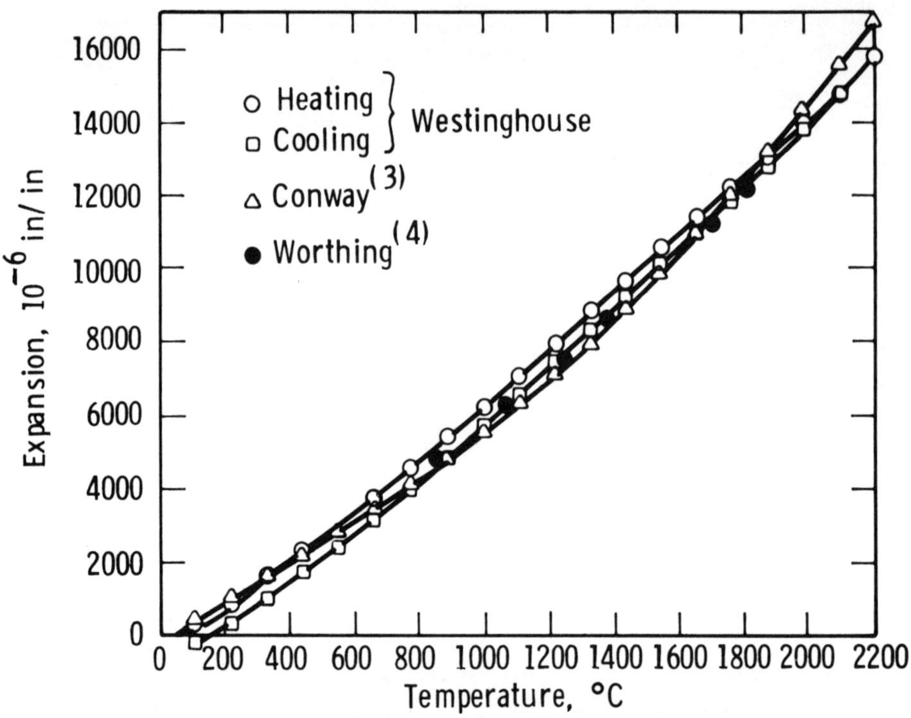

FIGURE 5–10.—Thermal expansion of arc cast molybdenum from 25°C to 2200°C to 25°C in vacuum.

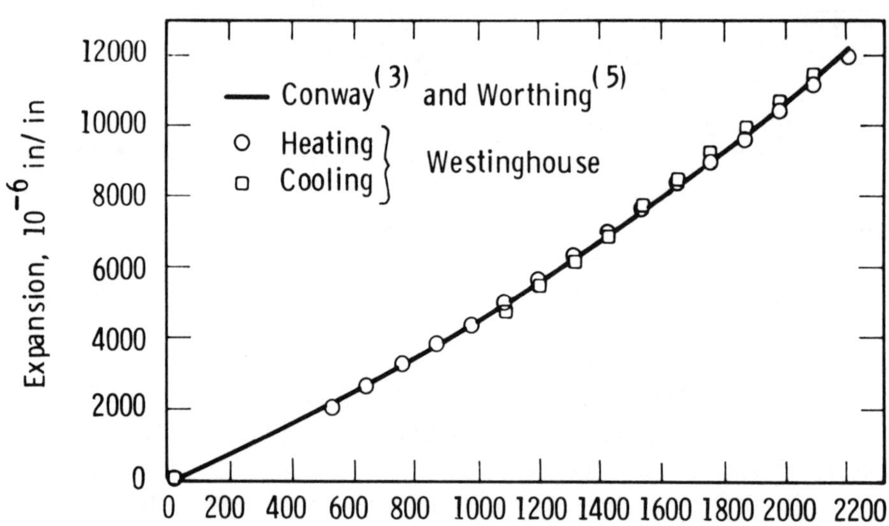

FIGURE 5–11.—Thermal expansion of AVC tungsten from 25°C to 2200°C to 25°C in vacuum.

obtained in this survey in both directions is about the same as that reported in the literature. Curves, however, are not presented in the literature.

Graphite deposits out on the thoria thermocouple insulators at high temperature in vacuum and the insulators become conductive, leading to erroneous temperature readings. For this reason, a new piece of insulator is used in the high-temperature zone for each test. This problem, which was encountered only in the graphite tests, was minimized by enclosing the thoria insulator in a tantalum tube. The thoria insulators can be used for several tests on metals before they become contaminated with metallic deposits.

It was found that Moseley Model 2A recorders, which do not have guard circuits, could not be used to record the data accurately. Because of the close dimensions of the tantalum heater, shields, and the tube, and the 7 kw power requirements; electrical pickup and feedback were recorded on the temperature axis of the recorder when a guard circuit was not used. With a guard circuit all electrical pickup and feedback were eliminated from the thermocouple signal. Figure 5-3 shows a Model 2A recorder, but the Model 7001 A, which has a guard circuit, is now used exclusively.

Items 10, 11, 12 and 31 in Figure 5-1 were included in the design to permit the use of optical or radiation pyrometers; but up to the present time, tungsten-3% rhenium vs. tungsten-25% rhenium thermocouples have been used exclusively.

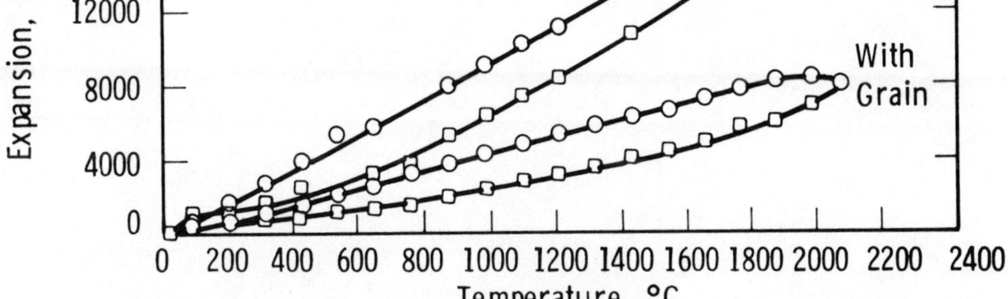

FIGURE 5–12.—Thermal expansion of ZTA graphite cross grain and with grain, 25°C to 2200°C to 25°C in vacuum.

References

1. J. W. Edwards, "High Temperature Structure and Thermal Expansion of Some Metals as Determined by X-Ray Defraction Data. I Platinum, Tantalum, Niobium, and Molybdenum," *Journal of Applied Physics*, Vol. 22, April, 1951, p. 24.
2. I. E. Campbell, *High Temperature Technology*, John Wiley & Sons, New York 1956, p. 50.
3. J. B. Conway, "Properties of Some Refractory Metals, III-Thermal Expansion Characteristics of Tungsten, Rhenium, Tantalum, Molybdenum, Niobium, W-25 Re., Ta-10 W, and MO-50 Re," GEMP-375, General Electric, Atomic Products Division, Cincinnati, Ohio, 1965.
4. A. G. Worthing, "Physical Properties of Well-Seasoned Molybdenum and Tantalum as a Function of Temperature," *Physical Review* 1966, Vol. 28, p. 190.
5. A. G. Worthing, "The Thermal Expansion of Tungsten at Incandescent Temperatures" *Physical Review* 1917, Vol. 10, p. 638.

CHAPTER 6

Special Dilatometer for Measurements on Plates in Vacuum

Introduction

On numerous occasions, there is a need for special tests which require the development of new equipment and test techniques. For example, special test equipment was designed, built, and used to measure the thermal expansion of Zircaloy plates. Since Zircaloy readily reacts with oxygen and forms an oxide which affects the measurement of its thermal expansion, a vacuum system had to be designed so that oxidation could be eliminated. Furthermore, the tests had to be run on large specimens.

Tests show that the apparatus is accurate within 3%. The system is also suitable for use with normal-size thermal expansion test samples and can be easily modified to test other odd-size samples.

Specimen

The Zircaloy plate section, 6 inches × 3½ inches × 0.140 inch, is shown in Figure 6-1.

Mechanical and Measuring Systems

The vacuum thermal expansion apparatus is shown in Figure 6-2, Figure 6-3, and Figure 6-4.

Two-2 inch steel channels, 19 inches long located 5 inches apart symmetrically with respect to the 4-inch center hole of the steel plate, are welded to the steel plate. Near the plate end two steel cross-pieces are welded between the channels. These cross-pieces are water cooled to protect the measuring transducer which they house. Conax fittings mounted in the steel plate are used to conduct the electrical wires from the vacuum chamber to the recording instruments.

FIGURE 6–1.—Zircaloy plate section.

FIGURE 6–2.—Vacuum thermal expansion test apparatus for plates.

High purity alumina insulators, ¼ inch in diameter with flanges, are fitted ¼ inch apart in holes in both webs of each channel, over a 6-inch length, at the free end. The flanges prevent the insulators from falling through the holes in the webs.

Tungsten wire, 0.035 inch in diameter, is threaded back and forth from one channel to the other around the insulators, forming a heater above and below the specimen, about ¾ inch from each face.

A fused silica tube and rod assembly is used for the expansion measurement. The specimen is placed at the front of the tube and contacted by a rod whose other end is spring loaded. The fused silica rod is fixed to the Statham transducer, which measures the thermal expansion. As the specimen expands, it pushes the rod; and this movement is measured by the Statham gage, whose output is a change in voltage. The fused silica tube is fixed to a steel housing which is fixed to the steel cross-piece between the channels. The Statham transducer, which is an unbonded wire strain element em-

FIGURE 6–3.—Zircaloy plate section mounted in vacuum thermal expansion test apparatus.

ploying a Wheatstone bridge circuit, is kept cool by shielding and by circulating water through copper tubing brazed to the cross-piece to which it is fixed.

Recording, Shielding, and Vacuum Systems

The thermal expansion as measured by the Statham transducer is recorded on the X-axis of a Hewlett-Packard *X–Y* recorder, while the temperature measured by a chromel-alumel thermocouple on the specimen is recorded on the Y-axis of the recorder. The temperature rise on the specimen is controlled with a Beck programmer

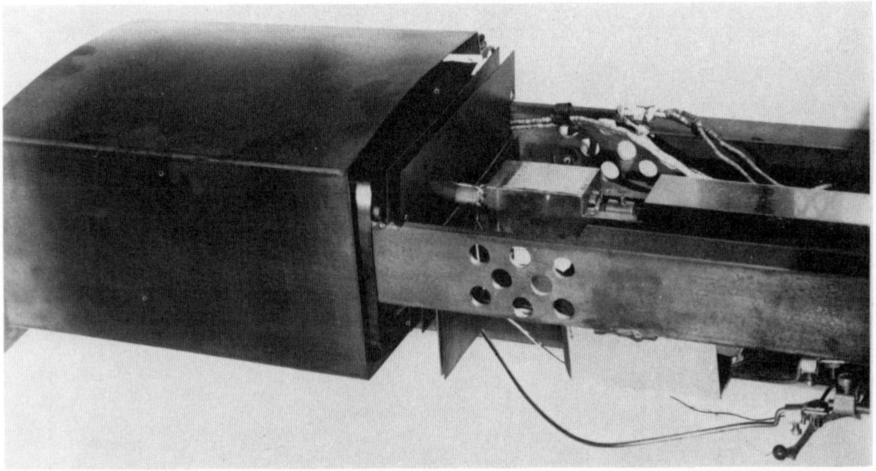

FIGURE 6–4.—Thermal shields around Zircaloy plate section.

through a second thermocouple on the specimen, and the temperature is recorded with a Bristol recorder.

The specimen and heater are shielded with a Zircaloy shield which in turn is shielded with a stainless steel sheet. Two shields were needed to keep the bell jar at a safe temperature. Additional shields were placed between the transducer and heater to minimize heat radiation to the transducer.

The system is enclosed with a Pyrex bell jar 30 inches long with an 11½-inch i.d. A wire mesh guard surrounds the bell jar. The vacuum seal is formed by a rubber gasket between the bell jar and the steel plate. The system is evacuated with a type 3304 NRC 4-inch high-vacuum pumping system.

Results

The thermal expansion data for a plain Zircaloy plate are given in Figure 6-5.

The results show that the expansion characteristics of the plain Zircaloy remained the same during two tests.

Because the temperature of the transducer could not be maintained at room temperature without drastically altering the system, it was necessary to calibrate the transducer under test conditions.

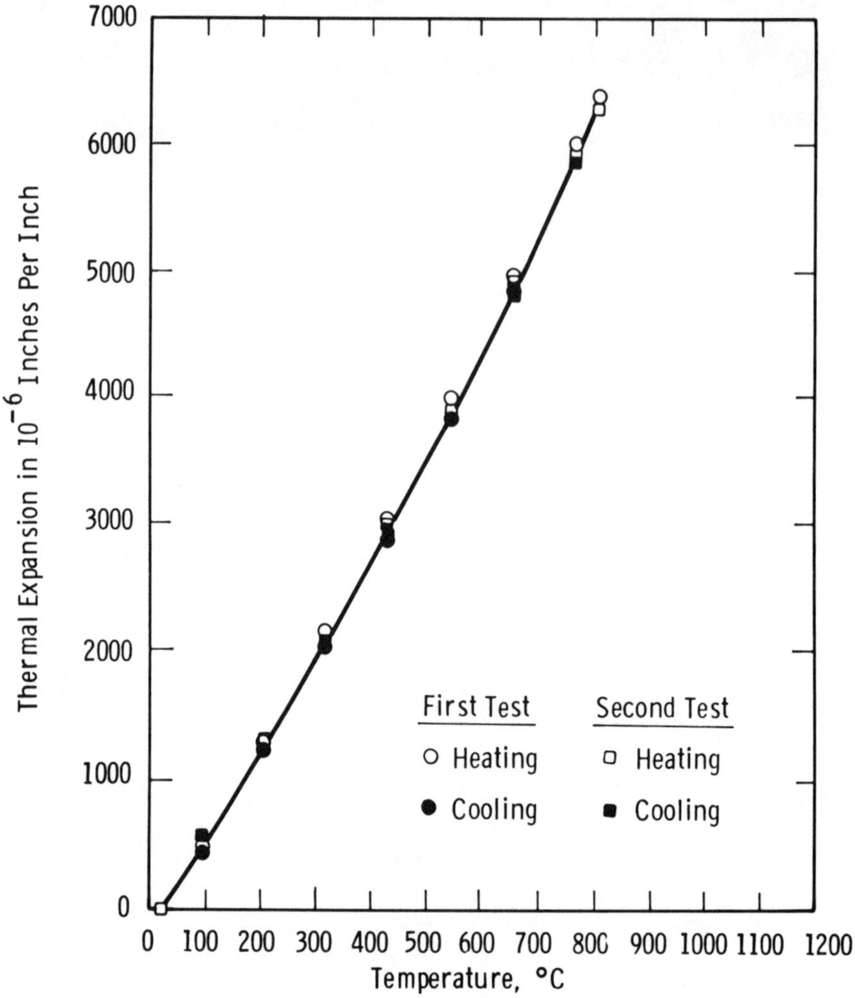

FIGURE 6–5.—Thermal expansion of pure Zircaloy in vacuum.

This was done by comparing curves made on a National Bureau of Standards specimen both in air and in the new vacuum system.

The results plotted in Figure 6-6 show that the secondary standard displays a higher apparent expansion when tested in the vacuum apparatus than when tested in air. The values obtained in vacuum are consistently 4% higher than those obtained in air. This factor was used as a correction factor for all vacuum tests made in this investigation.

Discussion

The apparently higher values of thermal expansion of the secondary standard when measured in the vacuum system were caused by the heating of the Statham transducer and by a reduced rate of heat dissipation during the test. Even though the transducer is water cooled and shielded from the specimen heater, its temperature still rises about 13°C at the maximum test temperatures; and this rise in temperature has the effect of increasing the output of the transducer, hence showing a greater apparent expansion. The specimen heating rate is kept constant from test to test; therefore the voltage output of the transducer resulting from heating is

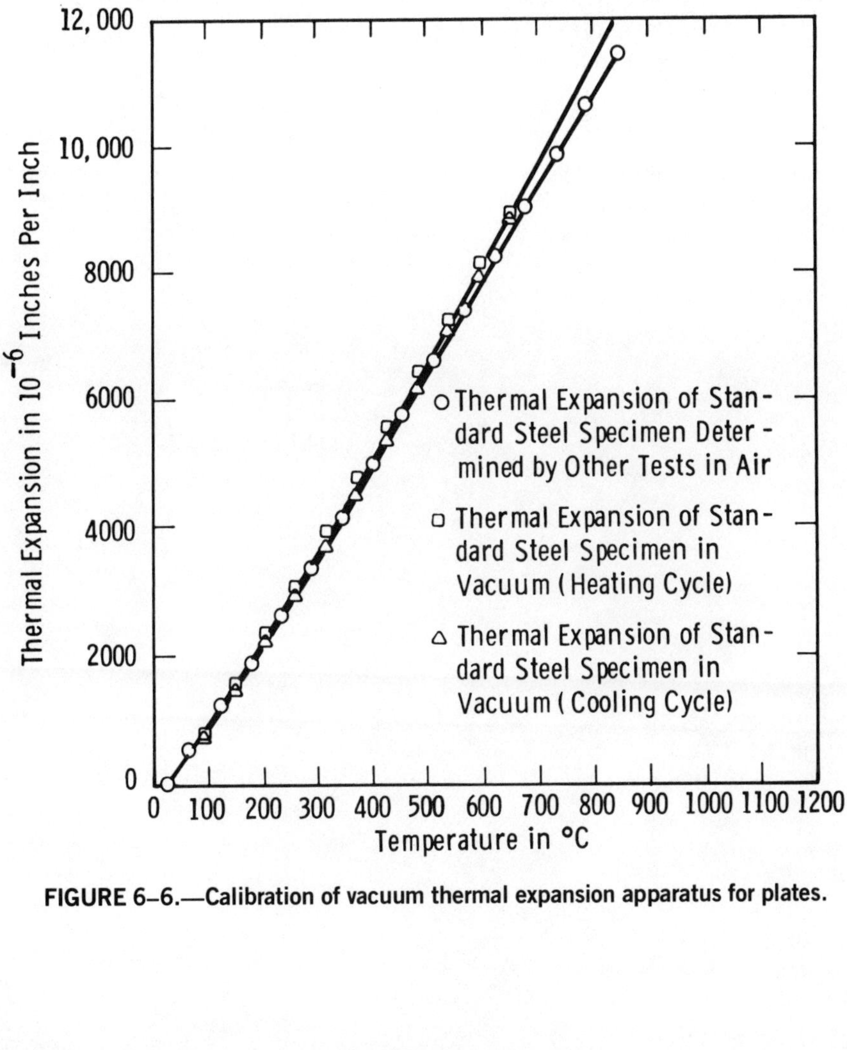

FIGURE 6–6.—Calibration of vacuum thermal expansion apparatus for plates.

repeatable. The calibration of the system using the secondary standard incorporated a correction for this effect.

The vacuum affects the Statham transducer by shifting the zero setting which is made at ambient atmospheres. However, after the vacuum reaches 6×10^{-5} mm of mercury, the Statham gage reaches an equilibrium condition and the zero setting remains constant.

The expansion values for plain Zircaloy obtained in this survey were compared with those obtained by other investigators who used x-ray methods, and the results of both methods agree well. The two runs made on the pure Zircaloy plate agreed within about 3%.

A nichrome wire heating element was first used in the system; but it vaporized, depositing metal on the aluminum insulators, resulting in grounding of the element. The nichrome was replaced with tungsten wire, which operated well. The tungsten wire does become very brittle on successive heating cycles, however, and care must be exercised to avoid breaking it.

The vacuum thermal expansion apparatus was designed to test large specimens but can also be used to test a standard specimen. It can be easily modified to test odd-shaped specimens of various sizes.

CHAPTER 7

Differential Dilatometry*

by W. A. Plummer

Introduction

Differential dilatometry is a method in which the thermal expansion of a sample is measured relative to that of a reference material. In conventional dilatometry the reference material is usually the measuring system itself, and this can be considered a special case of the differential method. However, the term differential is reserved herein for those cases where the reference material can be varied from test to test.

In many cases only the expansion of one material relative to another is required rather than the absolute expansion of either material. Differential dilatometry yields that value directly rather than

*This chapter on differential dilatometry is a paper titled "Differential Dilatometry—A Powerful Tool," by W. A. Plummer of Corning Glass Works, Corning, New York. Dr. Plummer is an expert in differential dilatometry. This paper appeared in AIP Conference Proceedings No. 17 (*Thermal Expansion*, 1973).

from computation of the differences from the absolute expansions. Some of the applications are in quality control, glass-sealing studies, strain measurements, and phase changes.

A primary advantage of the method is that a single test can suffice to determine the difference in expansion between two materials. The ability to match closely the thermal expansion, thermal mass, and thermal diffusivity of sample and reference specimens leads to further advantages. These include an increase in precision, a reduction of the errors due to temperature and extensometer calibration, the capaility of heating and cooling at high rates and of recording at high sensitivity. The method does not preclude the determination of absolute expansions, provided the expansion of the reference is known.

Method

Two designs of differential dilatometers have been employed in our laboratory. The first is a vertical type described at the First Expansion Symposium[1]. A schematic of this apparatus is shown in Figure 7-1. A 2-inch-diameter platform is suspended from three 0.25-inch-diameter support rods. The platform and rods are made from Corning Code 9454 material, a low-expansion glass-ceramic. The expansion of sample and the reference are transmitted through ball bushings by means of two push-rods. The differential expansion is measured by means of a differential transformer (LVDT), G. L. Collins model SS-101. One rod supports the coil of the LVDT with the core suspended from the other rod by means of an adjustable invar holder. The temperature of the LVDT is controlled about 20°C above ambient using a resistance winding and an inexpensive on-off controller. A counter-balanced resistance-wound tube furnace can be raised around the dilatometer assembly.

The second type of differential dilatometer is a commercial model available from Theta Industries. This is a horizontal unit constructed of a vitreous silica tube and push rods. Their Diaflex module provides independent, frictionless linear motion of either core or coil, the relative motion being measured by a Daytronic Model 201C differential transformer. Temperature control of the measuring system is accomplished by circulating a fluid from a temperature-controlled bath. A split-tube furnace, which can be rolled away from the dilatometer unit, is used.

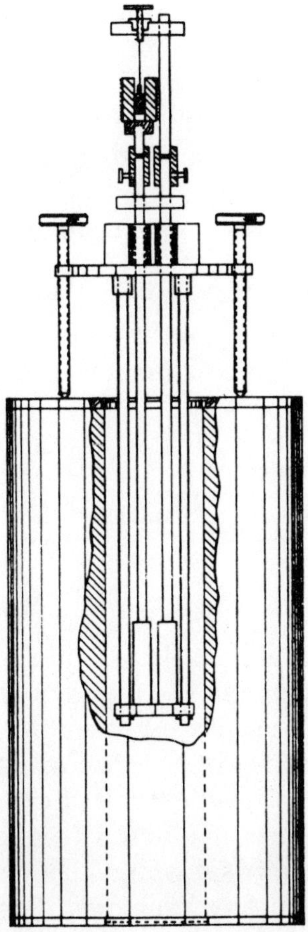

FIGURE 7–1.—Differential dilatometer, Corning design.

Calibration

In an ideal differential system the measured expansion is simply the difference in expansion between that of the sample and the reference. However, resulting primarily from differences in expansion between the two push rods, there is a contribution to the measured expansion, which we term a baseline correction. It is the apparent expansion which would be obtained if two identical samples were measured. The equation expressing the reltion between these expansions is

$$\Delta L/L_{\mathrm{smpl}} - \Delta L/L_{\mathrm{ref}} = A \times \Delta L/L_{\mathrm{meas}} - \Delta L/L_{BL} \qquad (1)$$

where A is a calibration constant and $\Delta L/L_{BL}$ the baseline correction, both of which must be determined.

Let us first consider the determination of the baseline correction, which can be accomplished by a pair of measurements on two samples. These samples may or may not have the same expansion, and they may have either low or high expansion. Both measurements should be made under the same experimental conditions, such as heating and cooling rates.

The first measurement will give equation (2) below

$$\Delta L/L_a - \Delta L/L_b = A \cdot \Delta L/L_1 - \Delta L/L_{BL} \tag{2}$$

where "a" designates sample and "b" reference. The positions of "a" and "b" are reversed in the dilatometer: "b" is now the sample and "a" the reference giving

$$\Delta L/L_b - \Delta L/L_a = A \cdot \Delta L/L_2 - \Delta L/L_{BL} \tag{3}$$

Adding these two expressions eliminates the two sample expansions and gives

$$\Delta L/L_{BL} = \tfrac{1}{2} A\,(\Delta L/L_1 + \Delta L/L_2) \tag{4}$$

In our laboratory the baseline is determined as a function of temperature over the entire range of the dilatometer. Three pairs of measurements are routinely used to establish this correction term. Four measurements are made using two specimens of Corning Code 9454 low-expansion ceramic. The other pair of measurements is between a Corning Code 9454 specimen and a Corning Code 7940 vitreous silica specimen. This latter set would also indicate the existence of extensometer or temperature problems. Baselines determined by this procedure have an uncertainty of 10 to 15 ppm based on a three-sigma limit. Quality control measurements on a large number of samples confirmed this uncertainty estimate.

The calibration of the extensometer can be accomplished in a number of ways. The method we prefer uses a precision screw such as a Pratt and Whitney measuring machine which has an accuracy of 10^{-5} inches. By measuring voltage differences over a range of 10^{-2} inches, one can attain accuracies of 1.0% or better. We then make use of two reference materials to verify the extensometer calibration. Again a pair of measurements is made in which sample and reference positions are reversed, the results being expressed by equations (2) and (3) above. Subtracting these equations yields

$$\Delta L/L_a - \Delta L/L_b = \tfrac{1}{2} A\,(\Delta L/L_1 - \Delta L/L_2) \tag{5}$$

For a calibration using 0° sapphire and Code 9454 ceramic the following values (ppm) apply at 900°C:

$\Delta L/L_a = 7622$	0° Sapphire
$\Delta L/L_b = 475$	Code 9454
$\Delta L/L_1 = 7204$	0° vs Code 9454
$\Delta L/L_2 = 7162$	Code 9454 vs 0°

Substituting in (5) and solving for A

$$A = \frac{(7622 - 475)}{\frac{1}{2}(7204 + 7162)} = 0.995 \tag{6}$$

The "best" value of A would be determined using data for the entire temperature range.

Error Analysis

A major advantage of the differential concept is the determination of differences in expansion with greater precision and accuracy. This is especially true when automatic recording is used. By choosing a reference with an expansion similar to that of the sample, one can greatly reduce errors due to uncertainties in the calibration of the extensometer and/or errors in recording. Table 7-1 shows what to expect for a calibration or recording uncertainty of one percent. The calibration errors begin to become significant only when differential expansions exceed 500 ppm. If the expansion difference can be reduced from 5000 to 500 ppm by the choice of references, the error from this source is reduced by an order of magnitude.

Table 7–1
Effect of a 1% Calibration Uncertainty

Expansion Difference ppm	Expansion Error ppm
0	0
100	1
1000	10
5000	50
10,000	100

A similar advantage accrues with errors traceable to temperature uncertainties. If one assumes an uncertainty of 5°C in the measurement of temperature from a recorder chart, the error in $\Delta L/L$ becomes significant only when the difference in expansion coefficient exceeds $1.0 \times 10^{-6}/°C$. Table 7-2 shows the errors to expect for various levels of differences in expansion coefficient between sample and reference. One effect of this is to permit recording with little or no loss of accuracy from errors in reading the temperature.

Table 7–2
Effect of a 5°C Temperature Uncertainty

Expansion Differential 10^{-6}/°C	Expansion Error ppm
0.0	0
0.2	1.0
1.0	5.0
5.0	25.0
10.0	50.0

By matching not only the thermal expansion, but also the thermal mass and thermal diffusivity, one can employ rapid heating rates with only a small sacrifice in precision and accuracy. Table 7-3 tabulates the maximum error based on what are believed to be reasonable assumptions, according to our experience. Reducing the heating rate by 50% decreases the error by a like amount. For maximum precision—and in cases where the expansion may depend on the cooling rate, as for glasses—slower rates should still be used. In many cases, rates of 450°C/hr. and 900°C/hr. can be used with satisfactory results. This is especially true where the absolute and differential expansions are small.

Table 7–3
Effect of Heating Rate

HEATING RATE: 900°C/Hr
ABSOLUTE TEMPERATURE ERROR: 20°C
RELATIVE TEMPERATURE ERROR: 2°C

Relative Expansivity 10^{-6}/°C	Differential Expansivity 10^{-6}/°C	Expansion Error ppm
1.0	0.0	2
5.0	0.0	10
10.0	0.0	20
1.0	0.5	13
5.0	0.5	21
10.0	0.5	31
1.0	1.0	24
5.0	1.0	32
10.0	1.0	42

Reference Materials

Two prime requirements for a good reference material are that its expansion be known and repeatable. Lack of stability of expansion on repeated cycling with differences between heating and cooling eliminates many promising materials from use as references. Secondary considerations are broad temperature range of use and similarity in expansion to the samples of interest both in magnitude and in temperature dependence.

The National Bureau of Standards has established a series of standards:

SRM 731 — a borosilicate glass
SRM 736 — copper
SRM 739 — vitreous silica

Single-crystal sapphire and platinum are also well-established reference materials.

Two other reference materials we have established for our use are a high-purity, high-density alumina and the previously mentioned Corning Code 9454 low-expansion ceramic. The alumina is used as a working standard in place of a single-crystal sapphire. Corning Code 9454 material has been a very useful reference for our work because it is usable over a wider temperature range than vitreous silica. It also matches more closely the expansion and temperature-dependence of many of the low-expansion materials of interest to Corning. This is important when accelerated testing is being performed, since high heating and cooling rates can be used. It also enables more accurate interpolations permitting removal of the furnace from around the dilatometer at temperatures as high as 300°C.

Calibration of Reference Materials

The differential dilatometer is an ideal instrument for establishing the expansion of new reference materials. For example, consider the calibration of an alumina reference by measuring its expansion relative to both the c-axis and a-axis of single-crystal sapphire. The alumina was run four times, twice with each sapphire sample. This procedure served to eliminate both the baseline correction and any uncertainties in the LVDT calibration. The following calculations illustrate the procedure:

Rewriting equation (5) for the alumina/sapphire expansion gives

$$\Delta L/L_{Al_2O_3} = \Delta L/L_{90} + \tfrac{1}{2} A\,(\Delta L/L_1 - \Delta L/L_2) \qquad (7)$$

$$\Delta L/L_{Al_2O_3} = \Delta L/L_0 + \tfrac{1}{2} A\,(\Delta L/L_3 - \Delta L/L_4) \qquad (8)$$

The following values (ppm) apply at 900°C:

$\Delta L/L_{90}$	$= 6874$	90° sapphire
$\Delta L/L_0$	$= 7622$	0° sapphire
$\Delta L/L_1$	$= +\ 93$	Al_2O_3 vs 90°
$\Delta L/L_2$	$= -\ 51$	90° vs Al_2O_3
$\Delta L/L_3$	$= -704$	Al_2O_3 vs 0°
$\Delta L/L_4$	$= +656$	0° vs Al_2O_3

"*A*" is determined by subtraction of (7) and (8) giving

$$A = 748/752 = 0.995$$

Inserting this value of *A* in (7) and (8) and adding yields

$$\Delta L/L_{Al_2O_3} = \tfrac{1}{2}\,[6874 + 72A + 7622 - 680A] \qquad (9)$$

$$\Delta L/L_{Al_2O_3} = 6946 \text{ ppm} \qquad (10)$$

If we assume an error of 1% in the calibration constant, the error in the total expansion for the alumina sample would only be 5 ppm or less than 0.1%.

This procedure illustrates another of the advantages of differential dilatometry: the ability, by two or more measurements, to verify directly the expansion of a sample. The calibration can be extended to an entire series of calibration materials in order to arrive at internally self-consistent values.

Applications

Quality Control

An important application for differential expansion measurements is for quality control. In fact, it was the necessity to control the expansion of ceramic materials for gas turbine heat exchangers that led to the first design and construction of a differential dilatometer in 1963. Expansion mismatch between components had to be controlled to within 100 ppm for the range 25°C to 900°C. The differential method permitted the direct measurement of the mismatch by one test. Moreover, faster heating and cooling rates could be used and the results recorded at a sensitivity of 100 ppm/inch. It is now used routinely for quality control measurements on gas turbine components, emission control substrates and on glass-ceramics for electric-range cooktops. For the latter application five values per day can be measured on heating for the range, 25°C–700°C. A large number of measurements yielded a one-sigma value for the variance of 20 ppm attributable to both specimens and measurement. Among six dilatometers used for this testing, only one showed a significant difference in the average expansion measured.

Glass-Metal Seals

An illustration of the differential expansion between a metal and a glass is that for Kovar® alloy vs. Corning Code 7052 glass. The main portion of Figure 7-2 shows the absolute expansion curves for both materials, while the insert shows the differential curve. Initially the expansion of the glass is lower than that of the metal. The expansivity of the alloy, however, decreases with temperature so that by 200°C the glass and metal have similar expansivities. From

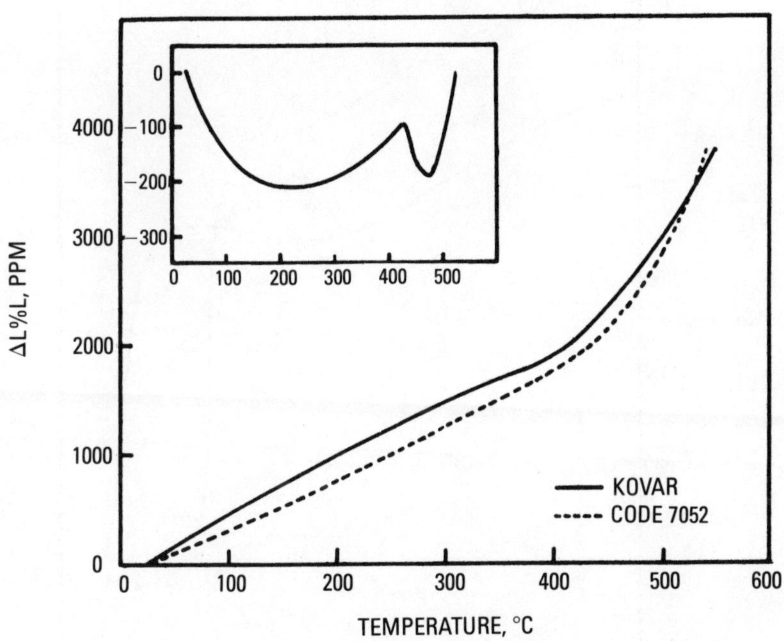

FIGURE 7–2.—Absolute and differential expansion curves for Kovar and Corning code 7052.

200°C to 400°C the expansivity of the glass is higher, and the differential becomes more positive. At 400°C Kovar undergoes a transition, the new phase having a much higher expansivity, thus causing the sharp drop in the differential curve. Above 460°C the glass expansion increases rapidly as the glass enters the transformation range. At 500°C the differential expansion becomes zero as indicated by the intersection of the two absolute expansion curves.

Phase Transformations

Phase transformations in ceramics and in metals can be readily measured with differential dilatometry. The temperature and magnitude of the change can be measured directly from the recorder chart. An example is shown in Figure 7-3 for a non-stabilized zirconia refractory.

Devitrification in glasses can be detected by the sharp increase in expansion at the transition temperature of the crystalline phase. Both cristobalite and quartz have been detected in this manner.

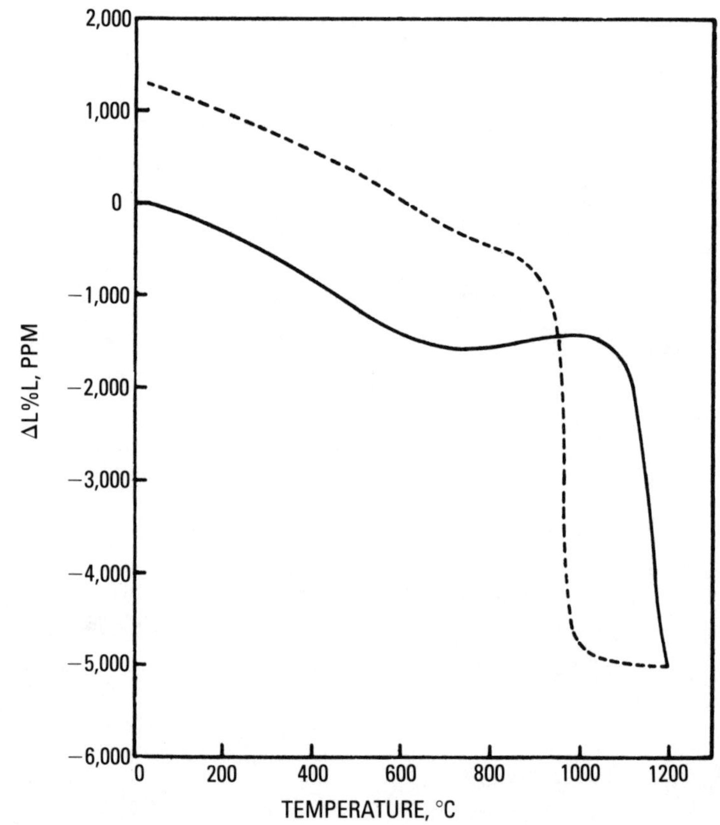

FIGURE 7–3.—Differential expansion curve: zirconia refractory versus alumina.

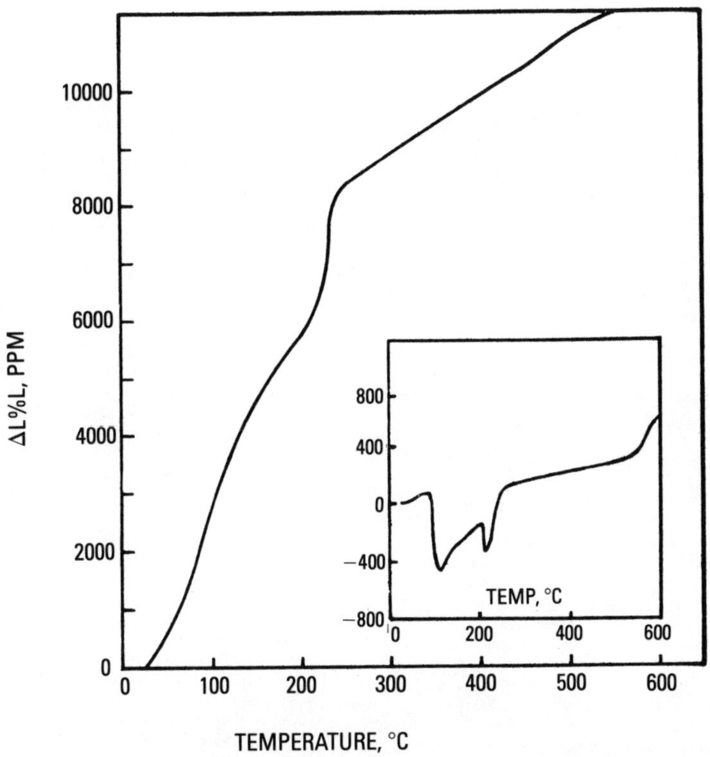

FIGURE 7–4.—Absolute and differential expansion curves for silica brick.

Vitreous silica dilatometers are subject to surface devitrification initiated by the presence of Na^+ at temperatures above 800°C. This is detectable by a break in the curve between 500°C and 600°C and is one of the limitations of a silica dilatometer.

A final illustration of phase transformations is shown in Figure 7-4 for two silica brick refractories. Three phases: tridymite, cristobalite, and quartz, are present, but in different amounts. The absolute expansion of one of the silica bricks is shown in the full curve, while the insert gives the differential trace. The one sample shows a smaller amount of tridymite as indicated at 100°C on the curve. The breaks at 230°C and 570°C indicate larger amounts of both cristobalite and quartz in this same sample. The same information can be obtained from absolute curves on each material, but the differential trace gives this information at a glance.

Reference

1. W. A. Plummer, "First Thermal Expansion Symposium on Solids," 1968, Unpublished.

CHAPTER 8
Commercial Dilatometers

The data given here on commercial dilatometers are presented objectively without favoring anyone's equipment. The most popular dilatometers of the four major manufacturers known to the author are presented. The reader should contact the manufacturers for any special equipment or information not given here. The information presented was taken from the manufacturers' latest brochures.

Harrop Dilatometers

The Harrop dilatometers are made by the Harrop Laboratories of Columbus, Ohio. The dilatometers are constructed horizontally using vitreous silica and alumina tubes for the specimen holders. The thermal expansion can be measured from – 180°C to 1600°C.

Harrop Model TDA-H1 Dilatometer

The Model TDA-H1 dilatometer is shown in Figure 8-1.

Principle of Operation

The sample is placed in a ceramic holder, in contact with a freely suspended ceramic probe rod. The holder and the sample are inserted in an electric furnace whose temperature is precisely varied and controlled by the control console. As the sample temperature changes and the sample expands or contracts, the change in length (ΔL) is transmitted by the probe rod to the core of a linear variable differential transducer (LVDT).

The LVDT converts this displacement into a proportional voltage signal which is recorded as the ΔL function. A thermocouple in contact with the sample provides the sample temperature (T) record. The resultant continuous plot of ΔL vs. T is the basic TDA curve.

Measurement Transducer, Sample Holders, and Furnaces

The suspension system and LVDT are mounted on a compound slide equipped with a precision micrometer head to position the probe rod conveniently and accurately with respect to the sample. These components, which comprise the measuring head assembly, are mounted on a low expansion invar support.

The measuring circuit provides a system linearity of 0.25% over a ΔL span of 0.2 inches. A continuously variable gain circuit with magnification up to 5000X permits direct, automatic recording of ΔL in convenient, even increments of the recorder for any sample length.

Standard models are available for three temperature ranges: 0-1000°C, 0-1200°C, and 0-1600°C. Fused silica sample holders and probe rods are used in the 1000°C unit; hi-purity alumina tubes are standard in the 1200°C and 1600°C models. Automatic, electronic

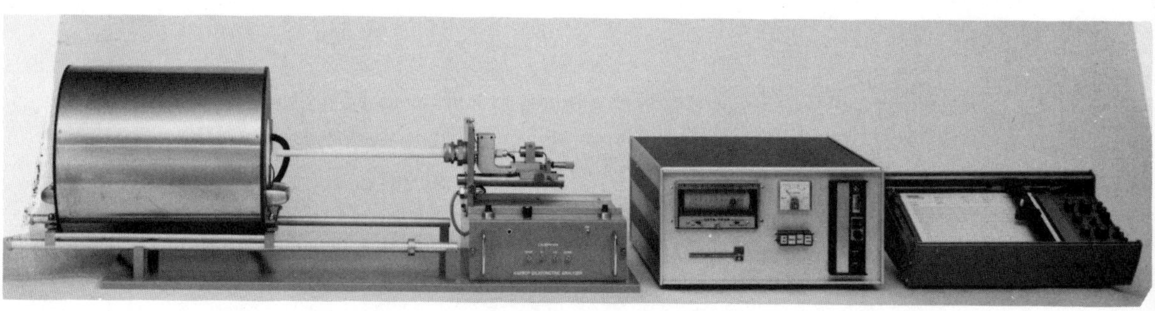

FIGURE 8–1.—Harrop Model TDA-H1 dilatometer.

correction for alumina expansion is a standard Harrop feature and is utilized when operating with the alumina tube. The TDA curves obtained are direct reading and do not require manual replotting to correct for alumina expansion.

The sample is heated by a metallic wound-tube furnace. The furnace is mounted on a moveable slide for convenient positioning over the sample. This feature provides free and easy access to the sample holder for sample placement and test set-up. The delicate measuring system is not disturbed, as the furnace simply slides over the sample holder to position the sample in the furnace hot zone area.

To provide uniform sample temperature, which is an absolute necessity for accurate dilatometry, all TDA-H1 Dilatometers incorporate a Harrop-designed, multi-shunt tube furnace, which provides better than ± 1°C/inch temperature uniformity over a four-inch hot zone. The helically wound, zoned heating element is a base-metal alloy in the 1000°C and 1200°C systems. A platinum alloy is incorporated in the 1600°C systems.

Programmer

CP Models—a mylar chart curve following the programmer provides both continuous and interruptable program capability for an infinite variety of heat, hold, and cool functions.

MP Models—A variable speed, motor driven programmer provides ten discrete linear rates of temperature rise and fall between 0.1°C to 15°C/min. Upper and lower temperature limit functions are used in conjunction with the following modes:

Rise to upper limit and OFF
Rise to upper limit and HOLD
Rise to upper limit and cool to lower limit

Temperature controller

A two-mode current proportioning controller operates with the furnace thermocouple to provide a continuously regulated closed-loop system. Control sensitivity is 0.1°C.

Power controller

Based upon system requirements, the power controller supplies variable power control from 0-95% of line voltage.

All instrumentation is mounted in an attractive metal enclosure. Necessary connectors for quick, accurate interfacing with the dilatometer module are included.

Recorder

The TDA-H1 system is supplied with a separate 8½ inch × 11 inch X–Y recorder. An optional 11 inch × 17 inch recorder is available.

Specifications

The specifications for the Model TDA-H1 dilatometer are given in Table 8-1.

Table 8–1
Specifications for Model TDA-H1 Dilatometer

Parameter	Range of Parameter
Temperature Range:	Model TDA-H1-CPO & MPO 1000°C Model TDA-H1-CP2 & MP2 1200°C Model TDA-H1-CP6 & MP6 1600°C
Temperature Uniformity:	±1°C/inch over 4″ Sample Length
Sample Size:	Standard Unit
Length	Any up to 4″ (100 mm)
Diameter	Any up to 7/16″ (11 mm)
	Large Sample Unit
Length	Any up to 4″ (100 mm)
Diameter	Any up to 1″ (25 mm)
LVDT Range:	±0.1″ (±2.5 mm) (Maximum Measureable ΔL (without repositioning LVDT)
System Linearity:	0.25% of LVDT Range
System Magnification:	0-5000X, Continuously Variable
Maximum Sensitivity:	Approx. 2×10^{-4} inch of ΔL/inch of recorder chart
Sample Atmosphere:	Controlled gaseous or vacuum to 10^{-3} torr (10^{-3} mm Hg)
System Overall Dimensions:	69″ W × 16″ D × 15½″ H (176 cm × 40 cm × 39 cm)
Weight:	85 lb. (39 kg.)
Power Requirements:	115 Vac, 15 Amps, 60 Hz.

Harrop Model TDA-H2 Cryogenic Dilatometer

The Harrop Model TDA-H2 cryogenic dilatometer is shown in Figure 8-2. It was specifically developed for studying the dimensional behavior of solids in subambient environments.

Chamber and Control System

The Harrop TDA-H2 dilatometer is comprised of two units: the cryogenic chamber and control system, and the main frame and measuring system common to all TDA-H Series dilatometers.

The cryogenic chamber is a double wall, stainless steel unit horizontally mounted on a movable slide for convenient positioning over the sample. The chamber consists of a helical wound heater, safety valve, cryogenic solenoid valve, and a gas circulator system which provides uniform distribution of the refrigerant over the sample.

The control unit consists of a programmer, SCR power control, ammeter, switching functions, and a dual function "ON-OFF" and "C.A.T." temperature controller. The "C.A.T." function operates the SCR power control, which powers the heating element; and the "ON-OFF" operates the cryogenic valve, which controls the refrigerant flow. This unique method of simultaneous control of heat and coolant, coupled with Harrop's gas circulating system, provides program control characteristics as well as temperature uniformity over the entire temperature range. Sample temperature is sensed by a thermocouple which can be placed near or in contact with the sample. Accessories for operation in controlled atmospheres or vacuum are easily attached.

The control package is available with either of two programmers. One is a variable speed, motor-driven type which provides program rates from 0.1°C to 20°C/min. Automatic shut down, hold, and re-

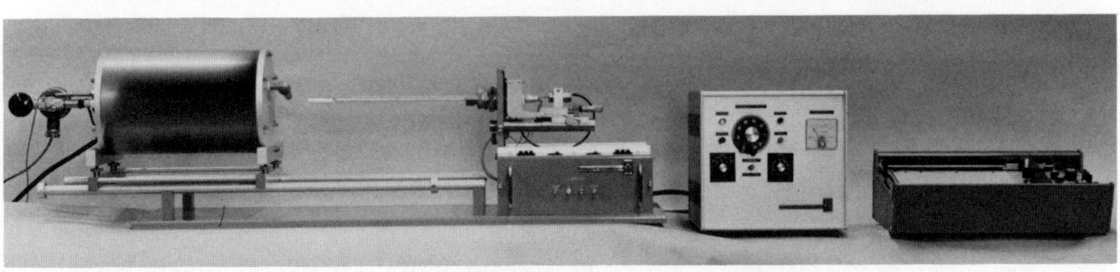

FIGURE 8-2.—Harrop Model TDA-H2 cryogenic dilatometer.

versing functions are included. The second is a mylar chart type which provides both continuous and interruptable programs for an infinite variety of heat, hold, and cool functions. This unit is ideal for simulating production processes. The TDA-H2 is available with either *X–Y* or Time-Base recording.

Principle of Operation

A sample is placed in a ceramic holder, in contact with a freely suspended probe rod. As the sample temperature changes and the sample expands or shrinks, the change in length is transmitted by the probe rod to the core of the LVDT transducer. The LVDT converts this displacement into a proportional voltage signal, which is recorded as the ΔL function.

High accuracy of the Harrop dilatometer is obtained by Harrop's unique design of the probe rod suspension system. It provides frictionless operation while applying minimal force to the specimen. The horizontal design eliminates gravity loading problems. The probe rod suspension system, LVDT, and micrometer head are mounted on an invar support to minimize errors caused by changes in ambient temperature.

This dilatometer utilizes an entirely new design concept for the calibration and amplification systems that enables it to handle samples of any length up to 4 inches. Fixed sample lengths are not required and calibration to any length is quick and easy. A ½-inch micrometer head with 0.0001-inch divisions positions the transducer for convenient zeroing and recorder calibration. The built-in amplifier has a continuously variable gain so that "percent of dilation" can be directly plotted in even increments on the recorder regardless of sample length.

Specifications

The specifications for the Model TDA-H2 dilatometer are given in Table 8-2.

Additional Features

The TDA-H2 is equipped with an electronic correction circuit for the recrystallized alumina probe and sample tube. Included also is the automatic dilation limit control system, which will shut down,

Table 8–2
Specifications for Model TDA-H2 Dilatometer

Parameter	Parameter Material or Range of Parameter
Probe Rod and Sample Holder	Fused Silica
Temperature Range:	−180°C to +500°C
Refrigerant:	Liquid Nitrogen
Sample Environment:	Air, Controlled Atmosphere, or Vacuum
Sample Size:	Length: 0.1″ to 4″, Width: 7/16″ Maximum
LVDT Range:	± 0.1″
System Linearity:	0.15% of LVDT Range
Magnification Range:	Completely variable 0-5000
Maximum Sensitivity:	Approx. 2×10^{-4} inch/millivolt
Power Requirements:	110 Volts, 600 Watts

hold, or reverse the temperature program should the sample being tested begin to melt or shrink below a preset point. This feature helps prevent damage to the dilatometer tubes caused by a melting sample.

The incorporation of these features and the use of the standard TDA-H Series main frame make the system completely adaptable for high temperature operation.

Netzsch-Geratebau GMBH SELB Dilatometers

The Netzsch-Geratebau GMBH SELB dilatometers are made in West Germany and sold in the United States by the Netzsch Brothers Inc., of Lionville, Pennsylvania. These dilatometers, hereafter called Netzsch dilatometers, are designed to operate in the horizontal and vertical orientation and use vitreous silica and alumina tubes to hold the specimen. They cover the testing temperature range − 160°C to 1700°C and operate with a common recording and control cabinet. Dilatometers are available for testing in air, gas, and vacuum atmospheres. A measuring head with a sensitive inductive transducer is used with the high and low temperature dilatometers.

Netzsch Model 402E Dilatometer

A diagram of the Model 402E dilatometer is shown in Figure 8-3. The various elements of the dilatometer are indicated in the Figure.

The measuring head shown in Figure 8-3 is used in all models except the 402EP and 402EV models.

The dilatometer measuring unit for high temperatures consists of a solid base plate on which the casing of the measuring head with the sample support and the furnace are arranged horizontally.

An airtight protection tube, which can be replaced by a protection tube with gas circulation covers the entire sample support. The furnace equipment rests on a carriage guide which is slid over the protection tube so that the test body is right in the center of the homogeneous temperature zone.

With the low temperature design the arrangement is vertical.

Measuring Method

The test specimen is located inside a tube which is closed on one side and has a recess in it. The tube is made either of vitreous silica (for temperatures up to 1000°C) or aluminum oxide (for temperatures up to 1700°C). A scanning rod made of the same material, supported free of friction, passes on the changes in length of the test body to an inductive transducer. The temperature of the test body is measured by a thermocouple.

Amplifier Device

The inductive transducer picks up the change in length, which is converted to a d.c. voltage and amplified. For amplification a carrier frequency measuring amplifier is used. The carrier frequency is 5 kHz.

Eight standardized measuring ranges are available:

25 50 125 250 500 1250 2500 500 μm
per 250 mm recording width.

In the most sensitive range this corresponds to an amplification of 10000. The automatic counter voltage on the recording apparatus permits an increase in the sensitivity of up to 100000 times.

Electrical Correction of Sample Length

A balancing circuit makes it possible to electrically standardize the measurement of test specimens between 25 mm and 50 mm long. This means that the test specimens are measured as if they

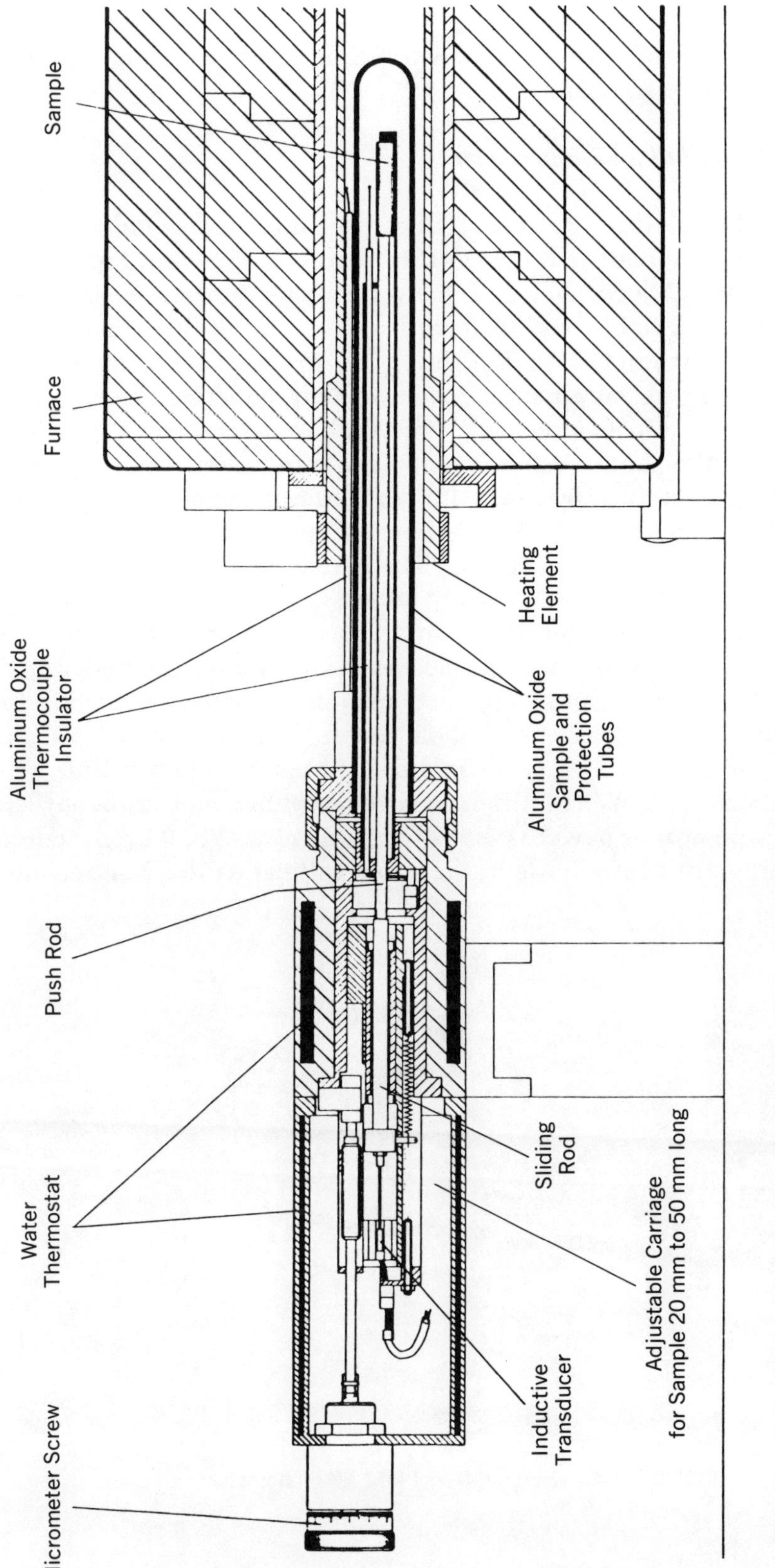

FIGURE 8–3.—Netzsch Model 402E high temperature dilatometer.

were all the same 50 mm long. With a series of measurements carried out on test specimens of varying lengths, the diagrams can therefore be easily compared.

Netzsch Model 402 EP Dilatometer

The Model 402 EP vitreous silica tube dilatometer for testing in air in the temperature range 25°C to 1100°C is shown in Figure 8-4.

The movable furnace and the dilatometer measuring head are mounted on a stable base plate with guide rails for the furnace. The furnace is wound with Kanthal wire for maximum service to 1200°C. The measurement is made with an inductive displacement transducer having a range of 5 mm. A micrometer screw is provided for specimen length adjustments and calibration. The specimen length is 25 mm to 50 mm with a maximum diameter of 12 mm. The data is plotted on an *X–Y* recorder, shown in Figure 8-5, using a paper size of 297 mm × 210 mm. The temperature is recorded on the *X*-axis of the recorder. Measuring ranges of 90 μm, 180 μm, 450 μm, 900 μm, 1800 μm, and 4500 μm are available on the *Y*-axis. The furnace temperature is controlled with a program controller with preselectable programs: up, up-constant, up-down, and constant heating or cooling rate selectable in ten steps between 0.1K and 100K per minute. Power is supplied with an SCR power unit with two silicon controlled rectifiers, control amplifier, and ammeter. The maximum output power is 3.5 KVA, from a 220 V, 50 cycle source. The program controller is in the same cabinet as the *X–Y*-recorder

FIGURE 8–4.—Netzsch Model 402 EP dilatometer.

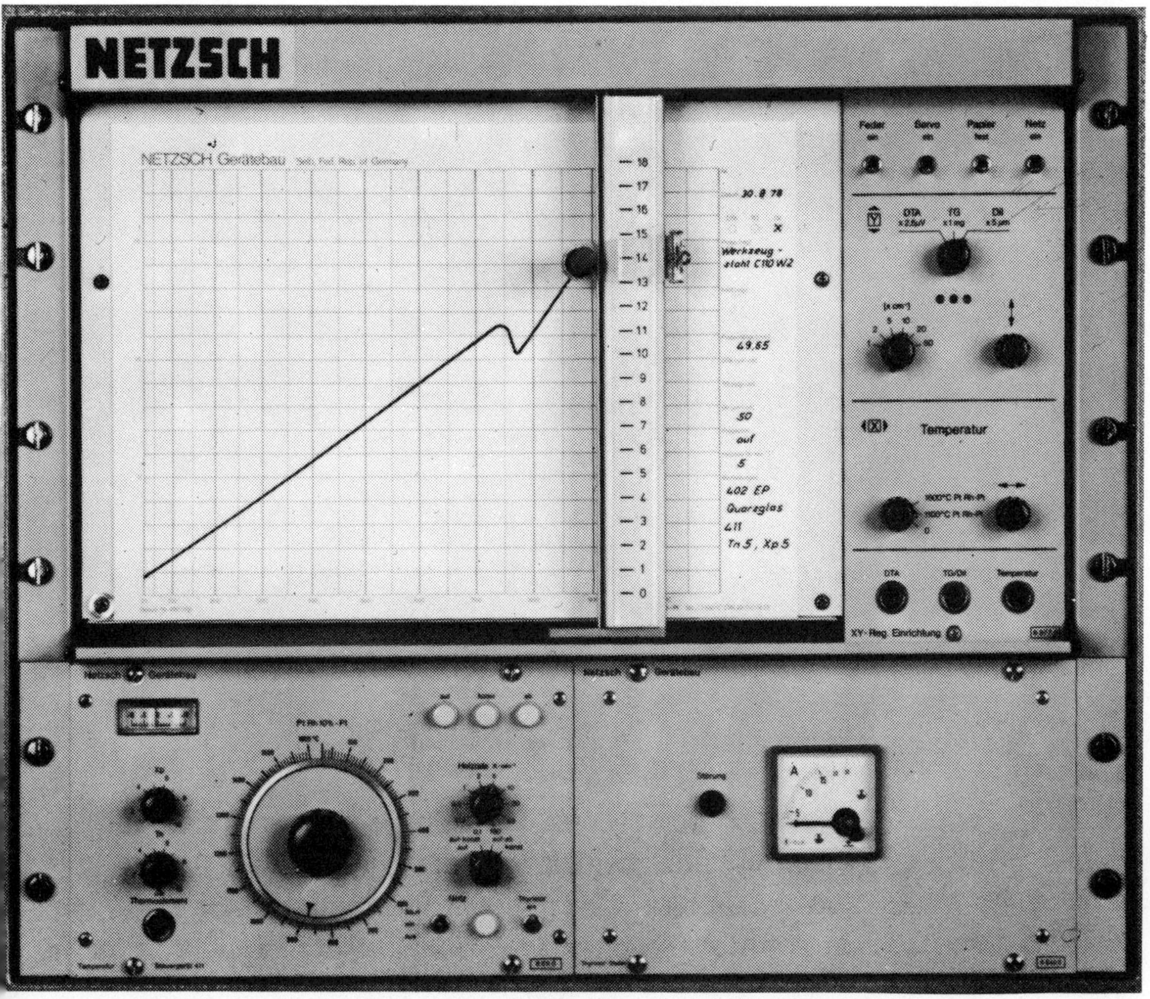

FIGURE 8–5.—Recorder and control unit for Model 402 EP Netzsch dilatometer.

shown in Figure 8-5. The dilatometer is 210 mm × 900 mm × 220 mm weighing 15 Kg, and the recording and control part is 435 mm × 565 mm × 452 mm and weighs 35 Kg. The specimen and control thermocouples are Pt./Pt.10% Rh.

Netzsch Model 402 ET Dilatometer

The low temperature dilatometer Model 402 ET with a furnace equipped with a bifilar resistance wire using a vitreous silica tube for tests in air, gas, and vacuum atmospheres in the temperature range −160°C to 420°C is shown in Figure 8-6. The maximum specimen size is 12 mm in diameter by 50 mm long.

FIGURE 8–6.—Netzsch Model 402 ET dilatometer.

Netzsch Model 402 ES Dilatometer

A high temperature dilatometer Model 402 ES which has a silicon-carbide tube furnace, an aluminum oxide specimen tube for testing in air, gas, and vacuum atmospheres in the temperature range 25°C to 1600°C is shown in Figure 8-7. The measuring head for this dilatometer is shown in Figure 8-3.

FIGURE 8–7.—Netzsch Model 402 ES dilatometer.

Netzsch Model 402 EV Dilatometer

The high temperature vertical dilatometer Model 402 EV, which is specially designed for large specimens for testing in air in the range 25°C to 1550°C, is shown in Figure 8-8. Specimens 35 mm in diameter and 50 mm long can be tested with this dilatometer. Specimens up to 20 mm in diameter can be tested under a preselected load.

Other Netzsch Dilatometers

Two other horizontal models, 402 E/6 and 402E/6R, which are not shown, are designed for testing in air, gas, and vacuum atmospheres from 25°C to 1700°C. They employ molybdenum or rhodium type furnaces and can test specimens 12 mm in diameter by 25 mm to 50 mm long in a gas or vacuum atmosphere.

A model 402 ED differential dilatometer, not shown here, is made by Netzsch. It is a horizontal vitreous silica tube type dilatometer for measurements from 25°C to 1100°C using a specimen 6 mm in diameter by 25 mm long.

A six-color recorder and control cabinet which can be used with Netzsch dilatometers is shown in Figure 8-9.

Orton Dilatometers

Orton dilatometers are made by the Edward Orton Jr. Ceramic Foundation of Columbus, Ohio. The dilatometers employ vitreous silica and alumina tubes oriented horizontally. Four models using vitreous silica tubes are available for operation to 1000°C. Model

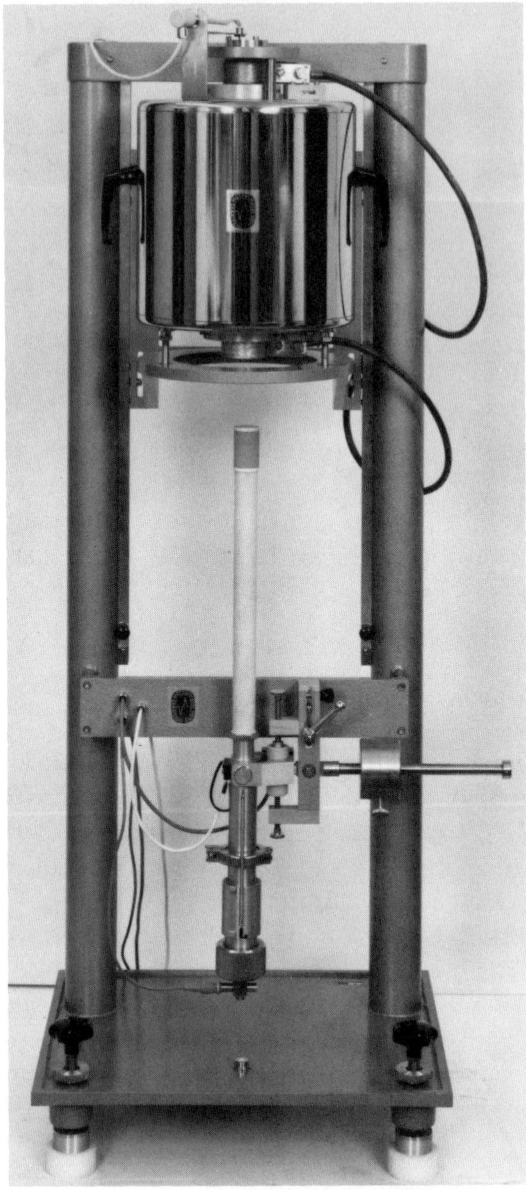

FIGURE 8–8.—Netzsch Model 402 EV dilatometer.

ES-1 has a single furnace, Model ES-2 has two furnaces, and both have a single full-scale temperature range of 1000°C. Model EK-1, which has a single furnace; and Model EK-2, which has two furnaces; have three full-scale temperature ranges of 250°C, 500°C, and 1000°C. One Model 15-E-02, using an alumina tube is available for tests to 1500°C.

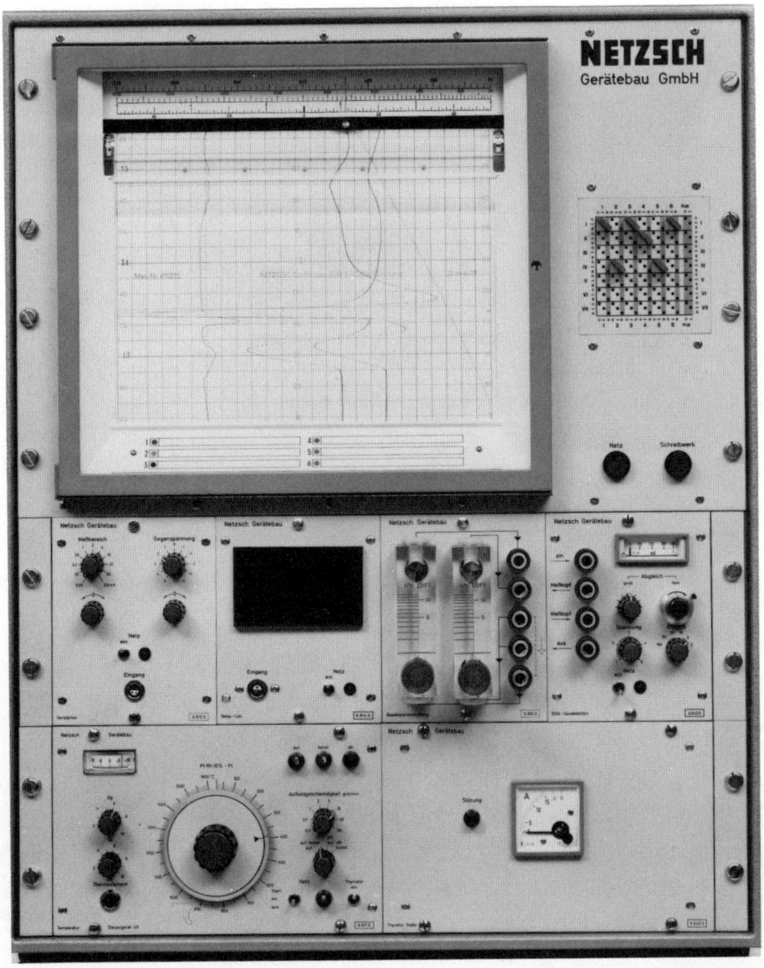

FIGURE 8–9.—Six-color printer and control cabinet for Netzsch dilatometers.

A block diagram of the Orton dilatometer is shown in Figure 8-10.

The sample rests in a tube (sample holder) within the furnace. The sample holder is rigidly connected to the body of the LVDT. The push rod rests in a sample holder and is in contact with the sample and the LVDT probe. Power is automatically fed to the furnace from the furnace power control at a rate according to the program setting.

The probe moves with the sample as it expands or contracts, displacing the LVDT core. This produces a voltage at the LVDT output that is directly proportional to the displacement. This voltage is amplified, and to it the sample holder correction factor is automati-

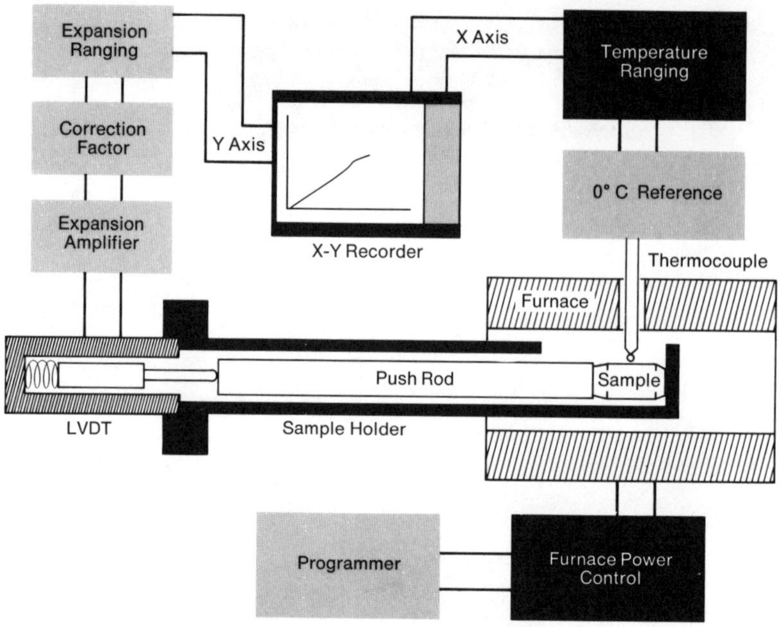

FIGURE 8–10.—A block diagram of the Orton dilatometer.

cally added. The expansion ranging control scales this signal for read-out on the Y-axis of the *X–Y* recorder.

The thermocouple is in contact with the surface of the sample and provides a voltage proportional to the sample temperature. This signal is referenced to zero degrees centigrade and plotted on the X-axis of the *X–Y* recorder.

Orton Models ES and EK Dilatometers

The Orton models ES and EK are shown in Figure 8-11. These models are physically identical but have different operational temperatures. The following is a description of the elements of the Models ES and EK.

A. An LVDT with a linearity of $\pm 0.1\%$ and a range of ± 0.1 inch is the heart of the expansion measurement system. To insure reproducible measurement of shrinkage, the transducer is helical spring-loaded to provide a uniform, gentle force of approximately two ounces throughout the linear stroke.

B. Precisely bored and ground surfaces on Orton sample tubes and pushrods insure stable and accurate translation of dimensional changes even in the most sensitive ranges. Standard sample length is two inches. However, an accessory kit is available to permit use of samples from .25 inch to four inches.

C. The Model S instruments utilize plug-in reference grade Platinum/Platinum-10% Rhodium thermocouples for greatest resistance to corrosion and for long-term stability at temperatures to 1000°C.

The Model K instruments use chromel-alumel thermocouples and have three linear temperature ranges. Automatic cold junction temperature compensation on all models insures accurate temperature plots.

D. Single-and-Dual-Furnace Models

Dual-furnace models are especially useful in laboratories with heavy work loads since they can practically double productivity. The second furnace permits another run while the first furnace is cooling. Dual-furnace versions are available in both the S & K series.

Conversion of single-furnace models to dual-furnace versions may be accomplished at a later date by means of a modification kit. Long-life Nichrome furnaces used on all models have temperature gradients less than ±2°C over the entire sample length during heating.

E. Operating controls provide the instrument with flexibility in programming and plotting. However, once the type of test is selected, a run may be started with only nine steps and in less than 10 minutes from removal of prior sample to the start of a new run.

F. Infinitely variable program controls permit a wide range of heating rates. Provision for fine adjustments allow studies requiring

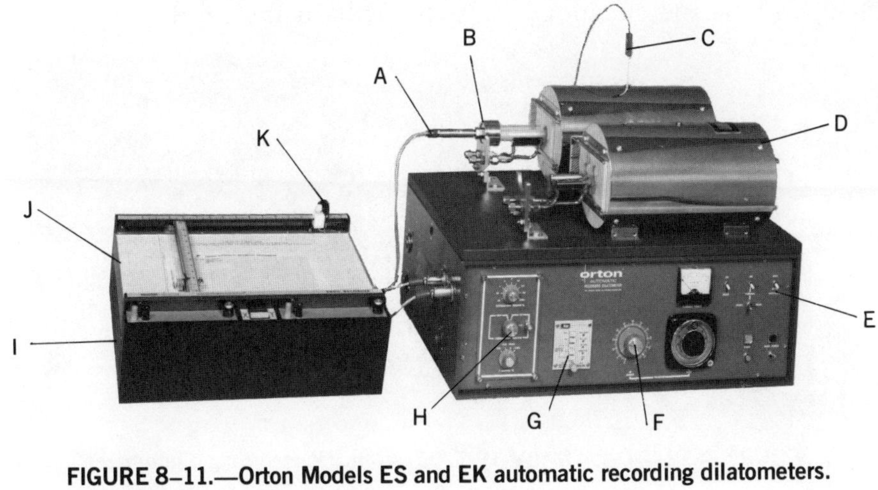

FIGURE 8–11.—Orton Models ES and EK automatic recording dilatometers.

very small changes in program rates. Programmed heat, hold, and natural cooling modes are provided.

G. The expansion signal conditioner contains controls for electrically setting the center of the linear range of the transducer and expansion range calibration adjustment.

H. Accurate dilatometry requires correction for the fused silica sample holder expansion over the sample length. The Orton dilatometer automatically supplies this correction and gives a direct reading of the true percentage of expansion or shrinkage. (Where coefficients of expansion are required, they can be easily calculated.) Eight switch-selected linear expansion ranges are provided by the precision Expansion Multiplier.

I. The variable temperature limit switch can be set at any temperature in the range. The unit automatically shuts off the instrument or starts the cooling plot.

J. The *X–Y* recorder is reliable, stable, and rugged. It accommodates either 8½-inch × 11-inch or 11-inch × 17-inch charts. (Units are delivered with a supply of 8½-inch × 11-inch graph forms, specially preprinted with temperature and expansion scales and title block.)

K. The shrink limit switch provides automatic protection against sample melts that might damage sample tubes.

The specifications for the Orton ES and EK models are given in Table 8-3.

Orton Model 15-E-02 Dilatometer

The Model 15-E-02 alumina tube dilatometer is shown in Figure 8-12, and the specifications for it are given in Table 8-4.

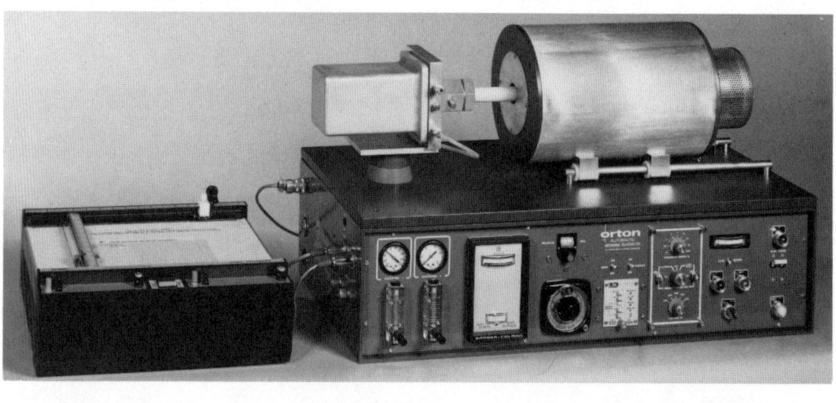

FIGURE 8–12.—Orton Model 15-E-02 automatic recording dilatometer.

Table 8–3
Specifications for Orton Models ES and EK Dilatometers

Dilatometer Element	Model ES-1	Model ES-2	Model EK-1	Model EK-2
Number of Furnaces	1	2	1	2
Thermocouple	Pt./Pt.-10% Rhodium Reference Grade (Type S)		Chromel-Alumel (Type K)	
Temperature Ranges	1000°C		250°C—500°C—1000°C	
Sample Size	Diameter: 0.95″ Maximum Length: 2″ (See Note)			
Accuracy (Laboratory Calibrated)	±½% of expansion of CP Platinum from 200°C to 1000°C in 0.7% range.			
Repeatability	¼% of full scale			
LVDT	Linearity ±0.1%			
Linear Range	±0.1″			
Expansion Ranges (Full Scale)	0.07%, 0.35%, 0.7%, 1.4%, 2.8%, 7%, 14%, 28%. Other or additional ranges available on request at time of purchase.			
Furnace	Nichrome Less than ±2° gradient over two inch sample length			
Program Modes	Programmed Heat, Natural Cool			
Heating Rates	To 25°C per minute			
Sample Environment	Air, inert atmosphere accessory soon to be available			
Sample Holder System	Fused Silica			
Readout	*X–Y* Recorder			
Power Requirements	115 volts, 10 amps, 50/60 Hz, Grounded			

NOTE: Units are normally laboratory-calibrated for two-inch sample lengths. They can, however, be supplied pre-calibrated for other lengths on request. (Accessory kits are available to permit use with samples from .25″ to 4″.)

Theta Dilatometers

Theta dilatometers are made by Theta Industries, Inc., of Port Washington, New York. They are the horizontal tube type and can be used to make measurements from – 150°C to 3000°C in air, vacuum, and gas atmospheres using vitreous silica, alumina, or refractory metal tubes. Several dilatometers representative of this line are described, including differential dilatometers.

Theta Model Dilatronic I Dilatometer

The Theta Model Dilatronic I dilatometer for measurements from 25°C to 1000°C is shown in Figure 8-13.

Table 8–4
Specifications for Orton Model 15-E-02 Automatic Recording Dilatometer

Dilatometer Element	Range of Element
Temperature Range	To 1500°C with Alumina Sample Holder System To 1000°C with Fused Silica Sample Holder System (Optional)
Expansion Range	Eight Switch Selected Ranges 0.14%-0.35%-0.7%-1.4%-2.8%-7.0%-14.0%-28.0% Full Scale (Expansion or Shrinkage)
Sample Size	1″ Long and up to ¾″ in Diameter Samples up to 2 Inches Long Can Be accommodated.
Accuracy (Laboratory Calibrated)	Calibrated against C.P. Platinum ±0.75%—Alumina Sample Holder System (400 to 1500°C) Optional ±0.5%—Fused Silica Sample Holder System (200 to 1000°C)
Repeatability	Plus and Minus One Half of Accuracy Specifications
Automatic Correction	Calibrated Accuracy Includes Automatic Correction for Sample Holder System.
Transducer (LVDT)	Linear Range: ±0.1 Inch Linearity: ±0.1% of Range
Thermocouple	Platinum vs. Platinum—10% Rhodium, Reference Grade
Furnace	Plug in-Furnace with Silicon Carbide Heating Element • Element Rated 1600°C • Furnaces Are Interchangeable—a Second Furnace Nearly Doubles Dilatometer Workload Capacity.
Program	Internal Modes: Heat—Cool to Room Temp. Manual Isothermal Hold Rates: Variable from 3°C/hr. to 6°C/min. Accessory: Accepts Data-Trak or Other Accessory Programmers
Sample Environment	Air or Controlled Atmosphere. Vacuum to Better than 1 mm Hg. Flow Controls Provided for Mixing Two Atmospheres
Power Requirement	230 V. 20 A. 50/60 Hz. (Grounded Service)

The dilatometer is a complete, self-contained compact system attractively styled in avocado green. It is fully automatic and requires no supervision during a test run. A split furnace allows easy access to the probe and fast cool-down between measurements. The famous Dilaflex Module allows entirely frictionless mounting of the pushrod. Two interchangeable modules, one for linear measurement and one for differential measurement, are available. Accurate calibration is attained with a built-in micrometer or precision gage block. The dilatometer has built-in automatic compensation for the expansion of the measuring system.

Technical Data

Temperature Operating Range	To 1000°C (1832°F)—150°C with Low Temperature attachment.

Automatically programmable with constant rate through motor driven indicating pyrometer (5°C/min). Three program modes: automatic increase, decrease, and hold

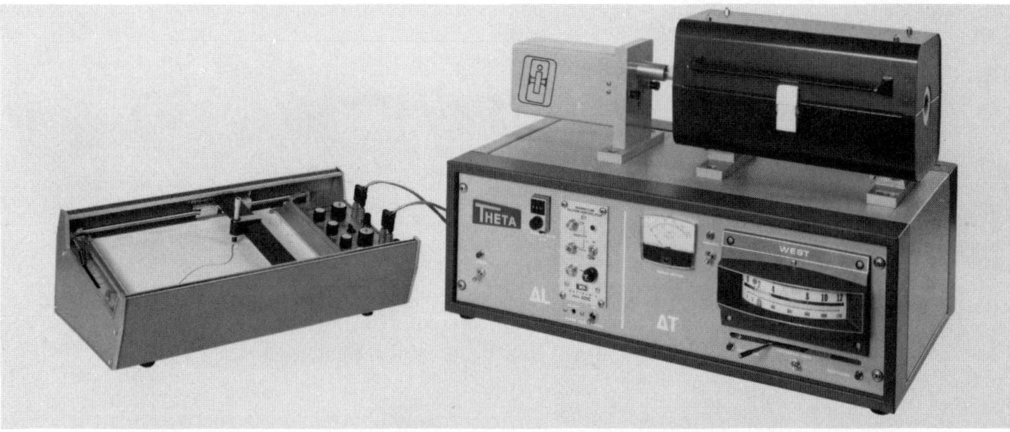

FIGURE 8–13.—Theta Model Dilatronic I automatic recording dilatometer.

Magnification range when used with 22cm × 28cm (8½″ × 11″) X–Y Recorder	10X-10,000X Selectable with a digital magnification control
Measuring System	Fused Silica
Sample Size	3mm to 60mm long 3mm to 12mm diameter ⅛″ to 2.5″ long— ⅛″ to ½″ diameter
Dimensions L×W×H	64cm × 41cm × 43cm 25½″ × 16″ × 17½″
Power Requirements	115 or 220 Vac 1000 Watts

Options

The modular design allows the system to be expanded with units such as DTA, constant load unit, protective atmosphere measuring head, and a variety of different measuring systems. Electrical resistivity measurement module, constant shrinkage programmer, low temperature attachment, and many other accessories are available.

Theta Models Dilamatic I and II Dilatometers

The Theta Dilamatic I and II dilatometers are shown in Figure 8-14.

Principle of Operation

The specimen to be measured is inserted into a sample tube in the center of the furnace. During heatup the expanding material

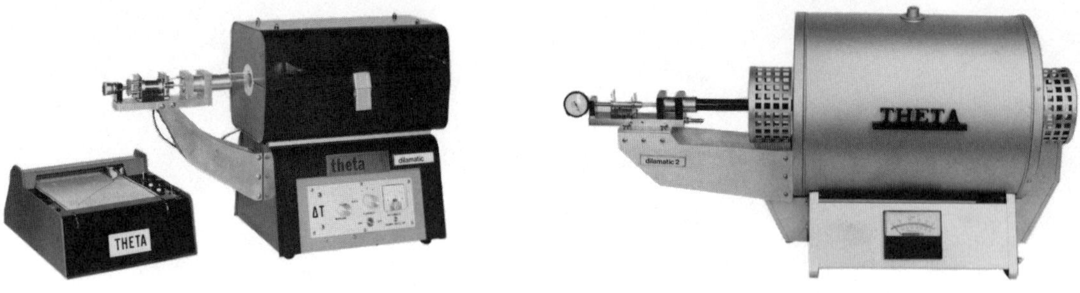

Dilamatic I Dilamatic II

FIGURE 8–14.—Theta Model Dilamatic I and II dilatometers.

activates a read-out device. A thermocouple attached to the sample measures the exact sample temperature.

The dilatometers as offered are complete and ready for operation. The automatic dilatometers require a recorder. Dilatometers are used to measure the thermal expansion of solids as well as phase changes in material under changing temperature conditions.

The need for an inexpensive thermal expansion apparatus for industry and educational institutions has brought about the new Dilamatic dilatometer series. The simplicity of the apparatus permits one to operate it without prior training. The rugged design lends itself to use in areas where scientific instruments normally cannot be found, such as production areas.

Depending upon the type of usage, either a model with a continuous read-out on a dial gage and an electric thermometer, as shown in Figure 8-15, or a fully automatic apparatus with automatic programming and recording is available.

The sturdy design of the measurement module does not have to be enclosed; thus the functioning of the apparatus is clearly visible at all times. This is of interest to the untrained user in industry and the student in a materials science course.

1. 0.0001 dial gauge.
2. Frictionless mounting spring.
3. Tube
4. Pushrod
5. Thermocouple
6. Sample
7. Constant temperature circulator or cooling water connection.
8. Thermocouple connection.

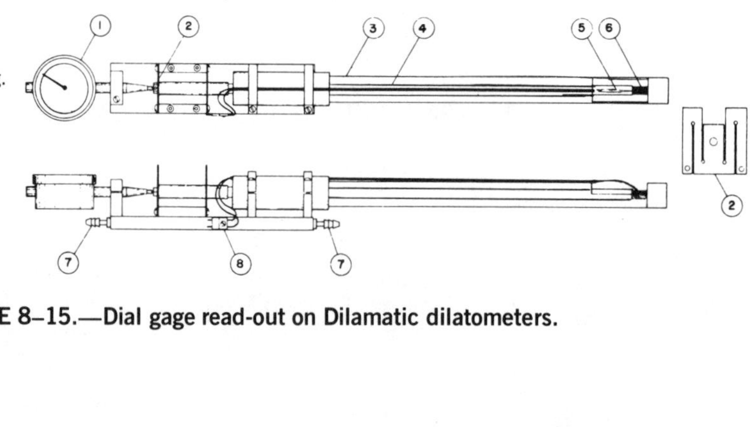

FIGURE 8–15.—Dial gage read-out on Dilamatic dilatometers.

Technical Data

Temperature Range	Room temperature to 1000°C (1800°F)—DILAMATIC I Room temperature to 1600°C (2900°F)—DILAMATIC II
Resolution	With dial gauge: 2.5 mm × 10^{-3} mm, 0.1 – 10^{-3} inch With 250 mm, 10″ strip chart recorder: 25× 10^{-5} inch
Measuring System	Fused Silica or sintered alumina
Sample Diameter	From 1 to 12 mm, 1/16 to ½″
Sample Length	1 mm to 60 mm, 1/16″ to 2½″ long
Power Requirements	1000 watts—115V or 220V—DILAMATIC I 3000 watts—220V—DILAMATIC II

Theta Model Dilatronic V Dilatometer

The Dilatronic V dilatometer is an ultrahigh temperature dilatometer for tests to 3000°C. A horizontal refractory metal tube is used to hold the specimen. The dilatometer is shown in Figure 8-16.

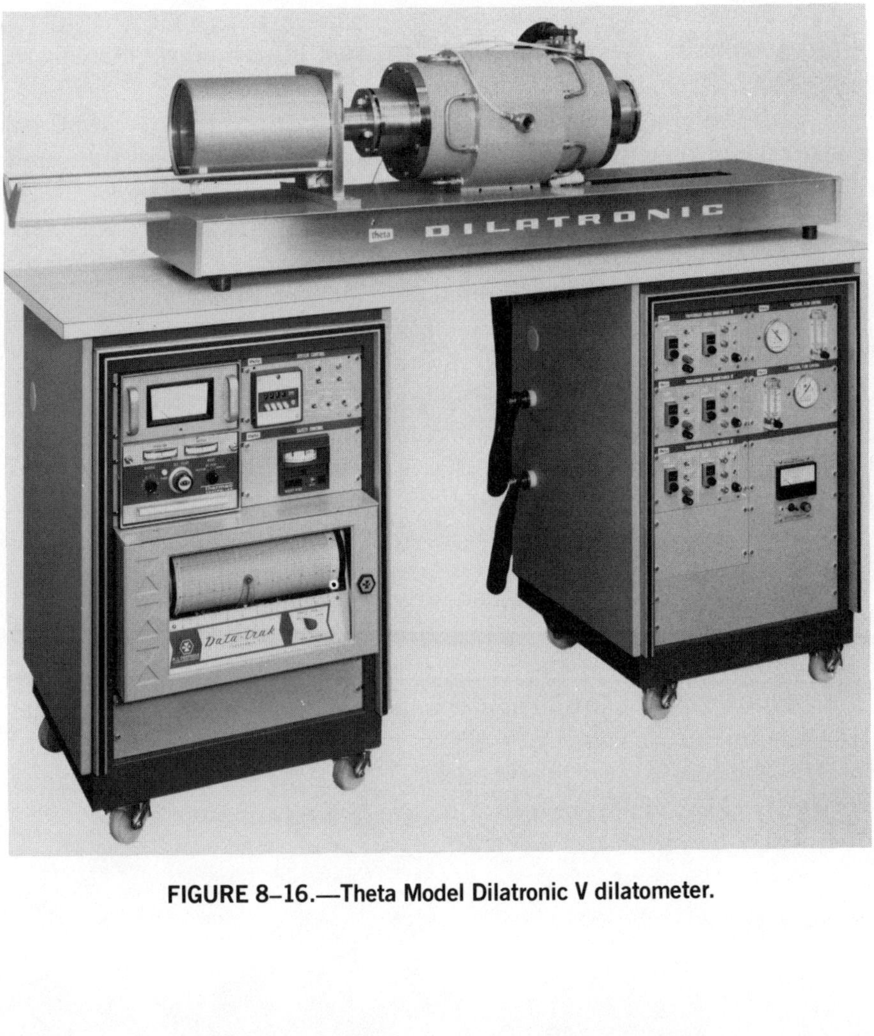

FIGURE 8–16.—Theta Model Dilatronic V dilatometer.

A floor model control console is used with a table top dilatometer. Single and double pushrod operation in vacuum or protective atmosphere is available. A digital recorder is also available for accurate readout.

Theta Differential Dilatometer

Elements of the Theta Differential dilatometer are shown in Figure 8-17.

With the Differential dilatometer, dimensional changes of a material are directly compared with each other or against a certified standard available from the National Bureau of Standards. The Differential dilatometer has a higher resolution than the standard single pushrod dilatometer.

Principle of Operation

Two samples, the reference and the test specimen are measured simultaneously with two pushrods.

There are two distinct movable elements: the frictionless mounted core assembly (Figure 8-17-1) and the frictionless mounted coil assembly (Figure 8-17-2). Both elements are held by two cantilevered springs (Figure 8-17-3) and assembled in interleaved fashion thus accomplishing the patented compact design (Figure 8-17-4). Both elements are displaced via pushrods by the length change of the specimen and the known reference material.

In operation the micrometer has to be disengaged to have the coil assembly free-floating. Item 9 in Figure 8-17-4 shows a screw adjustment that allows the LVDT core to be positioned relative to the coil. This adjustment compensates for possible slight differences in the length of the two samples.

Theta Dilatometer Measuring Head

The Theta dilatometer measuring head elements is shown in Figure 8-18.

The head is self-contained and usable for making measurements over a temperature range of −250°C to 3000°C under various atmospheric conditions and in a variety of configurations. The various features and principles of operation are given in Figure 8-18.

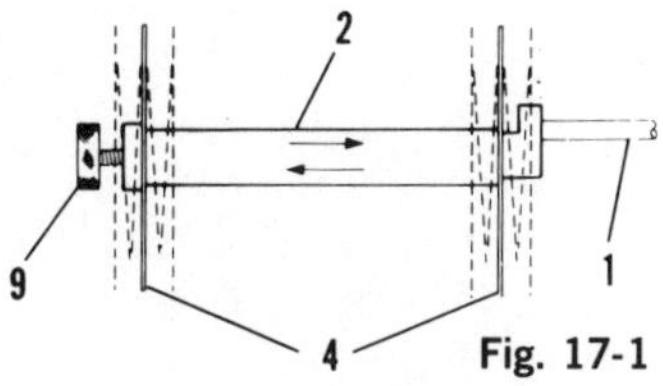

Fig. 17-1

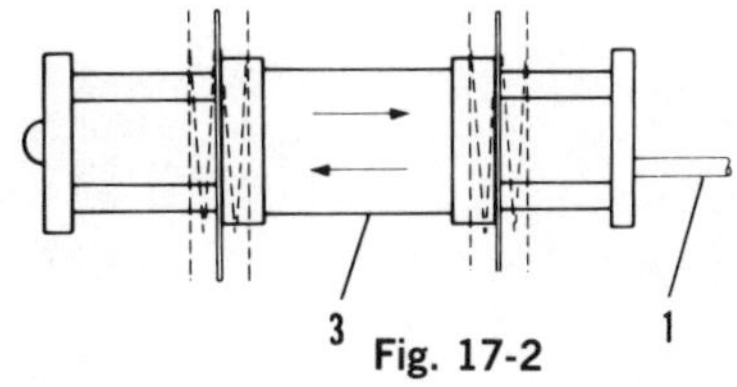

Fig. 17-2

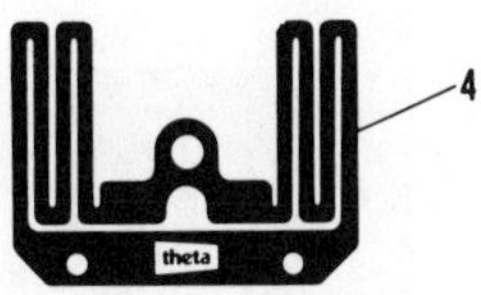

Fig. 17-3

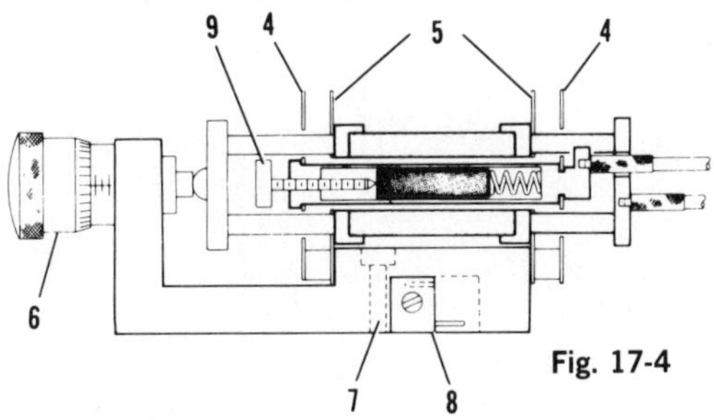

Fig. 17-4

1. Pushrod
2. Transducer core
3. Transducer coil
4. Pushrod springs
5. Coil springs
6. Micrometer
7. Mounting screw
8. Electrical connector
9. Core zero adjust

TECHNICAL DATA:
Length 110 mm, Height 48 mm, Depth 54 mm
Linear range ±1.25 mm, 0.25%
Force range 5 to 50 g
Elec. input 3V, 2.5KHz
Elec. output 500 mV/mm, 2.5 KHz

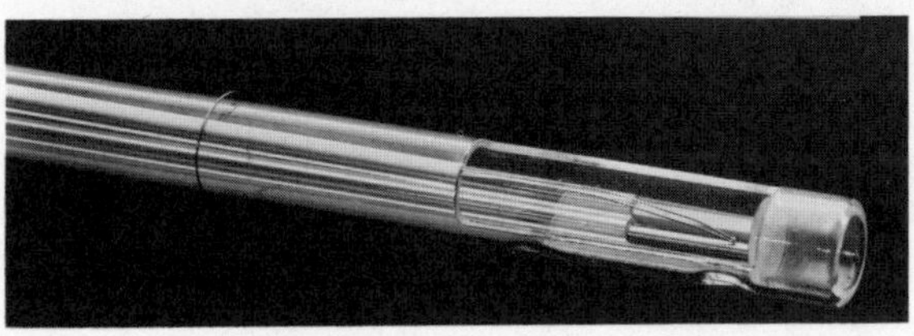

FIGURE 8–17.—Elements of the Theta Differential dilatometer.

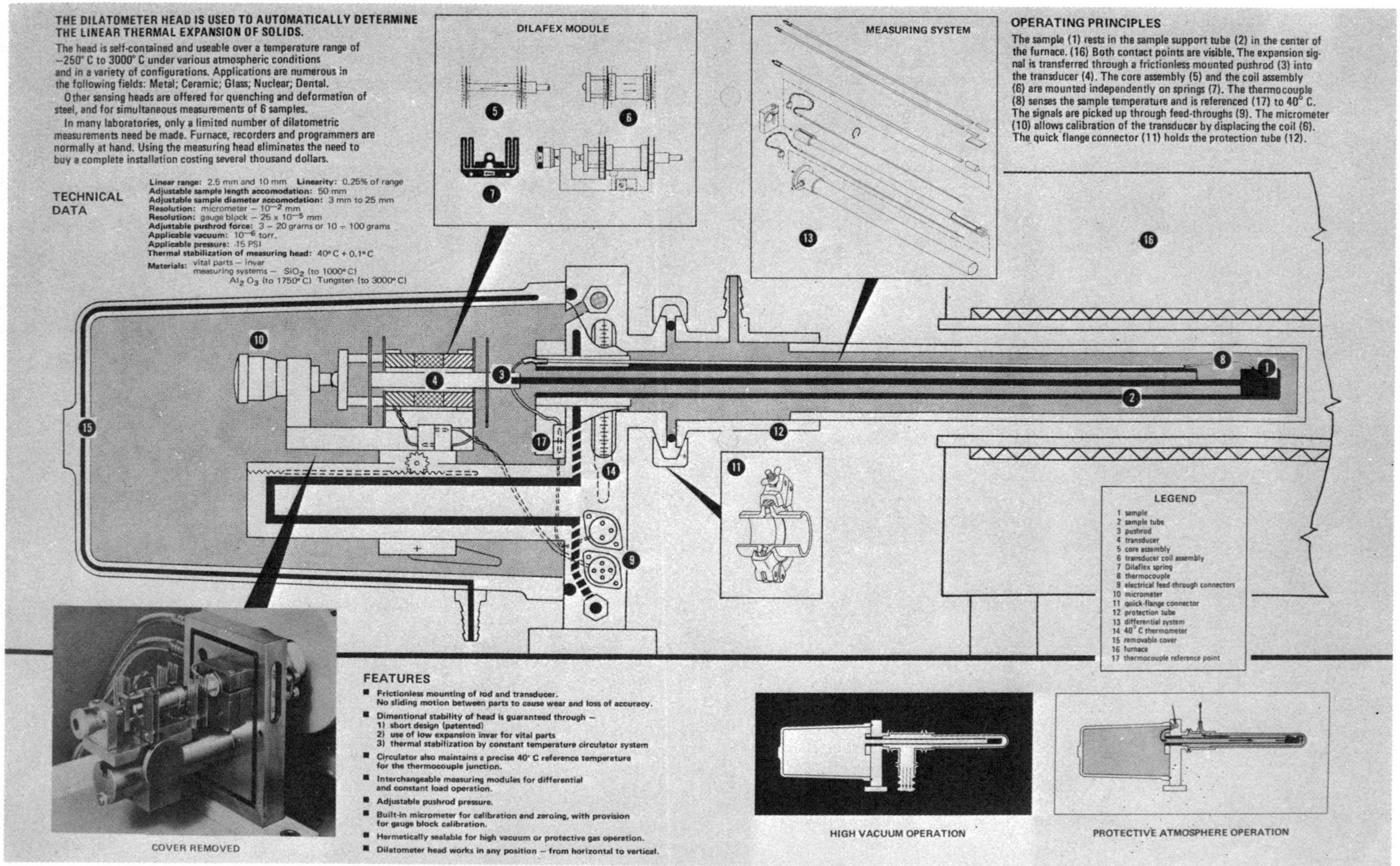

FIGURE 8–18.—Theta dilatometer measuring head elements.

CHAPTER 9

Specimen Size and Preparation Effects

Effect of Specimen Size, Preparation, and Heating Rate

Introduction

The specimen size, specimen preparation, and specimen heating rate were varied in a manual vertical vitreous silica tube dilatometer and a horizontal semiautomatic vitreous silica tube dilatometer. A Tuckerman optical gage was used to measure the expansion with the manual dilatometer, and a resistance wound-furnace was used to heat the specimen. The vitreous silica tube assembly with the Tuckerman gage head used with the manual dilatometer is shown in Figure 2-1 of Chapter 2. The semiautomatic dilatometer is described in Chapter 3. A Statham resistance-type transducer shown in Figure 3-6 of Chapter 3 and a Hewlett-Packard *X–Y* Recorder were employed with the semiautomatic dilatometer to measure and record the expansion. All tests were run to 650°C.

Experiment

The results presented here are not highly conclusive, since a limited number of tests were conducted. They will, however, serve as indicators and give the reader a better appreciation of the variables investigated.

A specimen ¼ inch in diameter by 4 inches long, supplied and calibrated by the U.S. Bureau of Standards, was used to determine the accuracy of both dilatometers. The effects of varying the specimen size, the heating rate, and the specimen preparation were determined by tests on a ¼-inch-diameter steel specimen. This specimen will hereafter be referred to as the survey specimen. The original tests were conducted on a specimen 4 inches long, and all shorter specimens were cut from this one. The thermal expansion properties of the 4-inch specimen were accurately established to 650°C in a previous thermal expansion survey. The rate of heating was regulated with a program controller. Normal care was exercised by the operators so that the results would represent an average test.

Four tests were run with the U.S. Bureau of Standards specimen to determine the accuracy of both dilatometers. Twelve tests were run with the survey specimen to determine the effects of varying the specimen size, the heating rate, and the specimen preparation.

Results and Discussion

Semiautomatic Dilatometer

The results, given in Table 9-1, show that the thermal expansion of the Bureau of Standards specimen was measured to an accuracy

Table 9–1
Semiautomatic Dilatometer Results

Specimen Length (Inches)	Identification	Specimen Preparation	Rate of Heating Specimen, °C/Min.	Range of Error %
2	Bureau of Stds. Specimen FC	Ends ground parallel With surface grinder	3.0 6.0	+3.2 −0
2	Survey Specimen	Ends ground parallel With surface grinder	6.0	+3.2 −0
2	Survey Specimen	Ends rounded with belt sander	6.0	+2.5 −1.1
½	Survey Specimen	Ends rounded with belt sander	6.0	+2.7 −0

of +3.2% −0% using a 2-inch-long specimen whose ends were ground parallel on a surface grinder. On the ½-inch-long and 2-inch survey specimens, with the ends machine ground or prepared by hand on a belt sander and with the heating rate as high as 6°C/min., an accuracy of +3.2% −1.1% was obtained.

A specimen 4 inches long could not be tested in the semiautomatic dilatometer because precision-ground vitreous silica tubes of suitable length were not available at the time of testing. Tests on a 4-inch specimen would be at least as accurate as those on the 2-inch specimen.

Manual Dilatometer

The results obtained with the manual dilatometer are given in Table 9-2. The thermal expansion of the Bureau of Standards specimen was measured to an accuracy of +2.2% −1.1% using a 4-inch specimen whose ends were ground parallel on a surface grinder. The 2-inch specimen was measured to an accuracy of +1.8% −1.7% and a 1-inch specimen to an accuracy of −7.7% +0%.

When a 4-inch survey specimen with machine ground ends was used and the heating rate was 2°C and 3°C per minute, an accuracy of +2.2% −1.9% was obtained. When the heating rate was increased to 6°C/min., the error increased. This increase in error is probably the result of a nonuniform temperature distribution over the specimen length, resulting from the rapid heating rate. When a 2-inch survey specimen was used, the error was decreased even though the heating rate was as high as 6°C/min. The 2-inch specimen probably had a better temperature distribution over its length than the 4-inch specimen.

The accuracy of the results is affected by the specimen length when the length is less than 2 inches. This effect is shown in the results obtained with the ½-inch and 1-inch specimens.

Conclusions

Semiautomatic Dilatometer

1. Specimens as short as ½ inch and as long as 2 inches can be tested accurately in the semiautomatic dilatometer.
2. The specimen ends can be ground parallel with a surface grinder or by hand with a belt sander.
3. The specimen can be heated as fast as 6°C/min. with no sacrifice of accuracy.

Table 9–2
Manual Dilatometer Results

Specimen Length (Inches)	Identification	Specimen Preparation	Rate of Heating Specimen, °C/Min.	Range of Error %
4	Bureau of Stds. Specimen FC	Ends ground parallel with surface grinder	2.0	+2.2 −1.1
2	Bureau of Stds. Specimen FC	Ends ground parallel with surface grinder	2.0	+1.8 −1.7
1	Bureau of Stds. Specimen FC	Ends ground parallel with surface grinder	2.0	−7.7 +0
4	Survey Specimen	Ends ground parallel with surface grinder	3.0	+2.2 −1.9
4	Survey Specimen	Ends ground parallel with surface grinder	6.0	+4.2 −1.4
2	Survey Specimen	Ends ground parallel with surface grinder	6.0	+1.7 −3.2
2	Survey Specimen	Ends rounded with belt sander	6.0	+1.8 −2.2
½	Survey Specimen	Ends rounded with belt sander	3.0	+19.4 −0
½	Survey Specimen	Ends rounded with belt sander	6.0	+14.5 −0
½*	Survey Specimen	Ends ground parallel with surface grinder	2.0	+3.4 −14.9
½	Survey Specimen	Ends ground parallel with surface grinder	3.0	+0 −30.8
½**	Survey Specimen	Ends ground parallel with surface grinder	2.0	+17.7 −2.6

*Vitreous silica tube holding specimen was raised 1½ inch above bottom of furnace.
**Specimen and furnace in horizontal position.

4. Testing with the semiautomatic dilatometer is less expensive than testing with the manual one.

Manual Dilatometer

1. Best accuracy is obtained in the manual dilatometer when the specimen length is 2 to 4 inches.
2. The maximum heating rate consistent with good accuracy is 3°C/min.
3. The specimen ends can be prepared either on a surface grinder or by hand, using a belt sander.

Ideal Specimen Size

As the result of many years of experience and the results of previous sections of this chapter, it is the author's judgement that the ideal specimen dimensions are ¼ inch in diameter by 2 inches long, as shown in Figure 9-1(A). The ASTM standards require a minimum diameter of 3/16 inch or 1/16 of the specimen length, whichever is smaller. This means that a 2-inch-long specimen could be ⅛ inch diameter, which is too small. With such a small diameter, it is difficult if not impossible, to make a thermocouple bead contact the surface of the specimen and stay in contact during the test. It could be argued that the thermocouple could be tied to the specimen with wire; however, with the techniques used with horizontal tube type dilatometers, tying the thermocouple with wire is not the best approach. Since the thermocouple wire is stiff, it could be attached to the specimen like a deflected contilever beam resulting in side loading that could move the specimen during the test. A second argument against a specimen ⅛ inch in diameter is that the specimen might buckle if heated to 1000°C in a vitreous silica tube dilatometer or to temperatures as high as 2200°C using the tantalum tube dilatometer described in Chapter 5. The ¼-inch-diameter specimen as opposed to a ⅛-inch-diameter or a 3/16-inch-diameter specimen gives assurance against buckling, and there is generally no difficulty in obtaining ¼-inch-diameter metal and ceramic specimens. All specimens should be the same size to maintain uniformity throughout the testing laboratories. If a specimen ⅛ inch in diameter or smaller must be tested, it should be placed into a loose fitting hole in a ¼-inch-o.d. cylinder as indicated in Figure 9-1(B).

Nonideal Specimen Size

There are times when it is impossible to obtain a specimen ¼ inch in diameter by 2 inches long. A specimen ¼ inch square by 2 inches long, shown in Figure 9-1(C), is satisfactory. The shortest length for good accuracy is 1 inch; below 1 inch the accuracy starts to suffer. A specimen shorter than 1 inch can be used if the dilatometer is calibrated with the same length standard specimen having about the same thermal expansion as the test specimen. Short

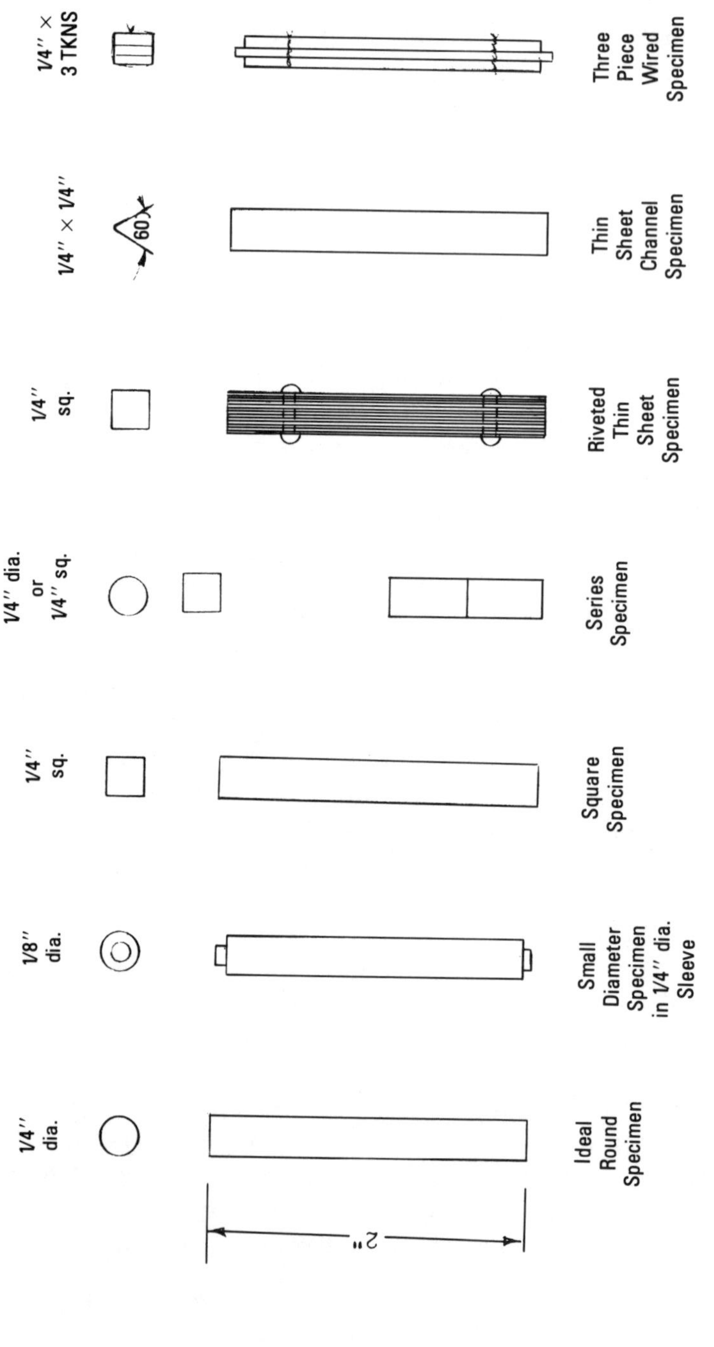

FIGURE 9–1.—Various specimen designs.

specimens can be placed in series in the dilatometer to form an acceptable length, as indicated in Figure 9-1(D), but the contacting ends must be parallel.

When one must measure the thermal expansion of a thin material, .010-inch-thick, for example, one can stack 25 pieces ¼ inch × ¼ inch × 2 inches and rivet them together as shown in Figure 9-1(E). If the maximum test temperature is not too high to cause buckling, one piece can be bent into a 2-inch-long channel specimen having a 60° angle as shown in Figure 9-1(F).

Tests on 1/16-inch-thick sheets of plastic material can be made on specimens cut ¼ inch × 1/16 × 2 inches with side plates of the same material ¼ inch × 1⅞ inches wired at both ends to the 2-inch specimen as shown in Figure 9-1(G). The measurement is made on the 2-inch specimen, and the side plates prevent buckling.

The change in diameter of a disc or ring specimen with temperature can be made by cutting out a through slot at the bottom of the vitreous silica tube as shown in Figure 9-2, and 9-3, and 9-4 respectively, and then inserting the specimen. With the ring, care must be exercised to make certain that the axial load holding the rod against the ring is small enough to prevent radial deformation due to loading. When larger specimens like the ring are tested, a furnace with a larger bore is necessary and the measuring part of the dilatometer requires no change.

Combining Dilatometer Components

Some materials operate in hot water environments and it is necessary to measure their thermal expansion in the operating envi-

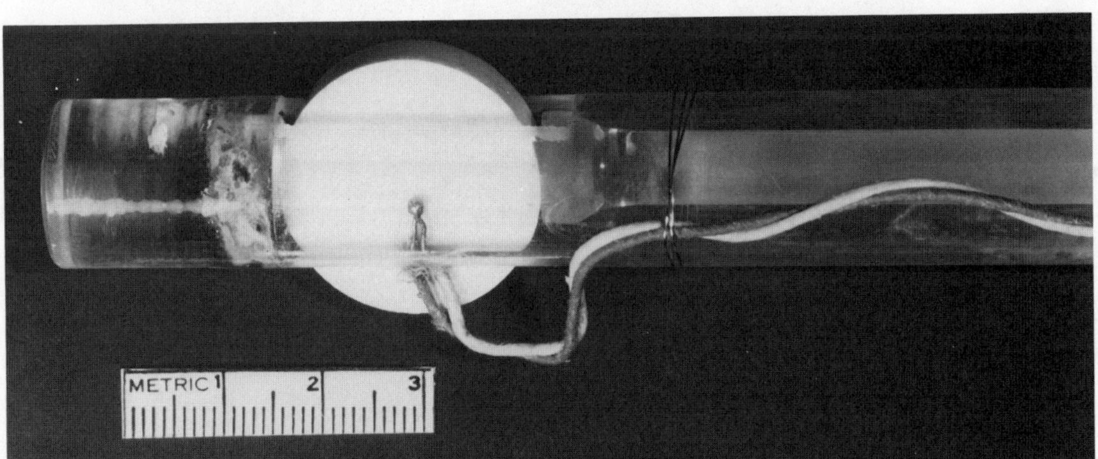

FIGURE 9–2.—Ceramic disc specimen, 1 inch diameter by ¼ inch thick, in vitreous silica tube.

FIGURE 9–3.—Vitreous silica tube dilatometer used to test 2 inches o.d. by ¼ inch thick ring specimen.

FIGURE 9–4.—Ring specimen, 2 inches o.d. by ¼ inch thick, in vitreous silica tube.

FIGURE 9–5.—Vitreous silica tube dilatometer used to test graphite in water.

ronment. This was the case for a graphite thrust bearing used in a canned motor pump. The thermal expansion of the graphite was measured from room temperature to the boiling point of water, using a specimen ¼ inch in diameter and 2 inches long. The stirred water bath described in Chapter 2 and the measuring section of the dilatometers described in Chapter 3 were used to make the measurements. The equipment used is shown in Figures 9-5 and 9-6.

FIGURE 9–6.—Graphite specimen in vitreous silica tube.

The transducer and vitreous silica tube pedestal and the steel rail base were turned vertically and bolted to the cantilever as shown in Figure 9-5. With this orientation the tube and specimen are immersed vertically into the stirred water bath. Since the bottom end of the tube at the specimen is open, the water flows around the specimen. To avoid turbulence around the specimen, a 2-inch copper pipe was placed around the tube. A close-up of the graphite specimen inside the tube after a test is shown in Figure 9-6. The sheathed chromel-alumel thermocouple is fixed to the vitreous silica tube rather than to the specimen because its stiffness makes it impractical to be fastened to the specimen. The specimen, in Figure 9-6, looks irregular because there is some water foam at the bottom of the tube, and the vitreous silica tube is not cut smoothly along the left edge. The bath temperature is raised at a rate of about 1°C per minute resulting in a uniform temperature throughout the bath. With this slow heating rate and good stirring action the thermocouple sheath and specimen are at the same temperature throughout the test.

An aluminum foil cover was kept over the bath to prevent water vapors and heat from reaching the transducer. The foil was arranged in such a manner in order to vent the water vapor radially outward. Tests with a secondary standard were conducted to establish a correction curve for the system. Tests on the graphite showed that the thermal expansion measured in water was about twice that obtained on dry specimens in an air atmosphere.

CHAPTER 10
Calibration of Dilatometers

Introduction

It is a common misconception that accurate thermal expansion results can be obtained with a vitreous silica tube dilatometer by correcting the data with handbook values for the thermal expansion of vitreous silica. The thermal expansion of vitreous silica can vary over a wide range as shown in the data in Figure 10-1. Even if one knows the precise thermal expansion of the tube material he is using, the most precise thermal expansion still cannot necessarily be measured because other system inaccuracies may exist. At high temperatures, the thermal expansion of vitreous silica is affected by foreign materials like the oils in human hands or by reactions with the material being tested. The foreign materials react with the vitreous silica and cause it to devitrify, resulting in a change of its thermal expansion. When vitreous silica is heated over about 1000°C, devitrification takes place even though foreign substances are not present, changing its thermal expansion. Thermal variations over the dilatometer can cause greater errors than the variations in the thermal expansion of vitreous silica. To obtain the most precise

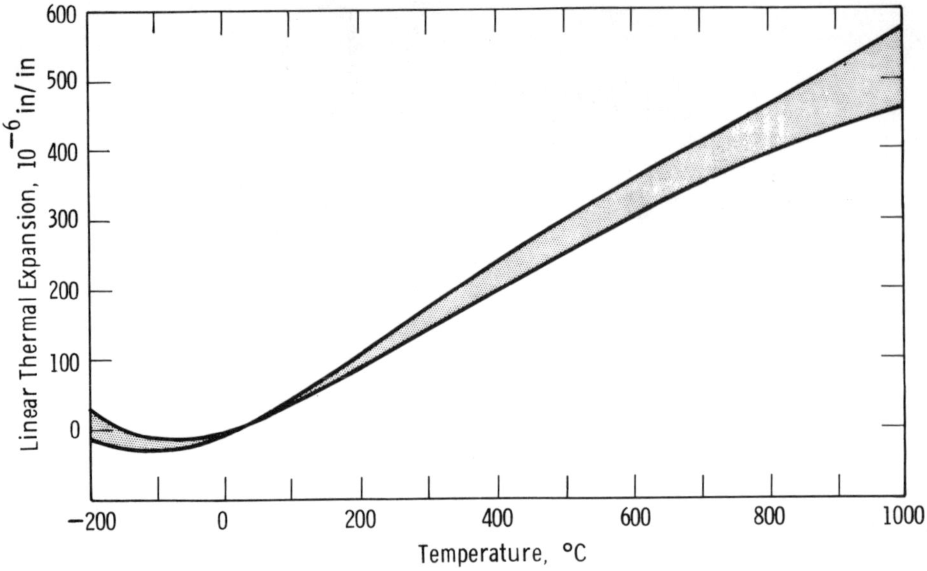

FIGURE 10–1.—Thermal expansion of vitreous silica.

thermal expansion of a material, the precise expansion of the tube does not have to be known at the time of testing, provided the system is calibrated with a secondary standard. The thermal expansion of tubes that are used over and over again, especially at higher temperatures changes so that repeated calibration is mandatory.

For optimum accuracy the vitreous silica tube should be handled with white gloves and periodic calibration performed with secondary standards.

Behavior of the Tube Type Dilatometer

Before proceeding to a discussion of calibrating a tube type dilatometer, one must point out several things for a better understanding of the discussions to come. The tube type dilatometer utilizes the comparative method of testing. The thermal expansion of a specimen is compared to the thermal expansion of the tube material; and after due consideration is given to the tube material and the dilatometer, the thermal expansion of the specimen can be determined. Therefore, the thermal expansion of the tube material must be known along with system performance.

In a tube type dilatometer the specimen is held at the bottom of the tube during the measurement. A rod made from the same material as the tube contacts the free end of the specimen, and as the specimen expands it pushes the rod. The movement of the rod is a measure of the thermal expansion of the material and is measured by a gage fixed between the tube and rod at the open end of the tube. This simple explanation assumes that tube and rod are at the same temperature so that the rod compensates for the expansion of the tube. At the bottom of the tube where the specimen is located there is no rod to compensate for the expansion of the tube, and it is for this reason that a correction must be made in the experimental data to take into consideration the expansion of the tube. The tube material should have as low a coefficient of thermal expansion as possible so that the correction is as small as possible, and the tube material must have the capability to withstand repeated heating to the top temperature of the test. The tube material must also be thermally stable so that the same thermal expansion occurs every time it is heated. It is desirable that the expansion be the same on heating and cooling so that the same correction values can be used during the heating and cooling cycle of a test.

Vitreous silica is an excellent material from which to make a tube for thermal expansion testing. It can be used to test materials from absolute zero to 1000°C in air over and over again, and its thermal expansion is comparatively small, typically 0.5×10^{-6} inches/inch °C in the range *RT* to 1000°C, the range in which most thermal expansion measurements are made. Above 1000°C in air, alumina is the accepted material for the tube and rod to about 1600°C, and above 1600°C one must test in a vacuum. For vacuum testing, tungsten is the best tube and rod material although tantalum has been used extensively by the author principally because a tantalum tube was considerably less expensive than a tungsten tube. Other tube materials could be molybdenum or certain grades of graphite and sapphire.

As mentioned previously, the thermal expansion of the tube compensates for the thermal expansion of the rod, but this compensation assumes that the tube and rod are at the same temperature. It is not an easy matter to keep the tube and rod at the same temperature during a test, but it can be done quite well, and any deviation from the ideal can be compensated for in a correction factor for the system. To emphasize the point; suppose the tube is at a higher temperature than the rod when a measurement is made. This means that the change in tube length will be greater than the corresponding change in rod length and a lower specimen expansion will be measured. If, on the other hand, the rod is at a higher temperature than the tube, a greater specimen expansion will be measured because the rod is longer than it should be. So it is imperative that the tube, rod, and specimen be at the same temperature when a reading is made.

It is also important to keep the measuring gage at the free end of the tube at the same temperature (generally at room temperature) during the test. Care must be exercised not to heat the measuring gage with the heat from the furnace, which would again introduce error into the expansion measurement.

Experience has shown that it is economically impossible to heat the entire tube and rod to the same temperature at which the reading is made. It is generally satisfactory to have the tube and rod at nearly the same temperatures along their lengths. In other words, a temperature gradient can exist over the length of the tube, but this gradient should be about the same on both the tube and rod. Any slight mismatch can be compensated for in the calibration of the system. That is why the system must be calibrated, and that is why the average thermal expansion of the tube material is not satisfactory as a correction factor to obtain the most accurate results.

It was stated earlier that vitreous silica is the best tube and rod material because of its low thermal expansion and good thermal stability to about 1000°C. It would be desirable to have tube and rod materials having similar thermal stability and low thermal expansion for use above 1000°C in air. Unfortunately such a material does not exist and one must be willing to give up either or both of these properties to conduct tests above 1000°C in air. Of course, optics can be used to make the thermal expansion measurements above 1000°C. Here, however, we are dealing with tube type dilatometers exclusively.

Alumina can be thermally stabilized and used in air to over 1600°C, but since the maximum limit of a platinum resistance furnace is about 1600°C, it is convenient to construct a dilatometer that is usable for testing to 1600°C. Sapphire could be used, but is not available in long enough lengths; and since alumina is less expensive, it has become accpeted as the standard tube and rod material to 1600°C. The thermal expansion of alumina is about 10.16×10^{-6} inches/inch °C in the temperature range 25°C to 1600°C, which is about 20 times that of vitreous silica and this is undesirable. As a result of this large thermal expansion, special care must be exercised with an alumina tube dilatometer to obtain accurate data.

Above 1600°C thermal expansion testing must be conducted in vacuum, and tantalum or tungsten tubes used to hold the specimen.

Tantalum tubes and rods, which have been used by the author because they are less expensive than tungsten tubes and rods, have a coefficient of thermal expansion of 7.8×10^{-6} inches/inch °C from 25°C to 2200°C. Since the thermal expansion of tantalum is also high, special care must be taken to obtain accurate results with the tantalum tube dilatometer.

Furnace

A resistance furnace, which is normally used to heat the specimen, must heat the specimen uniformly so that the temperature over the specimen length is the same when the thermal expansion measurement is made. This is best accomplished by using a multi-tap furnace or one that has several zones that can be independently controlled. The multi-tap type furnace made by Marshall and others is suitable.

To make certain that the furnace will heat the specimen uniformly one must make a temperature profile of the furnace, using thermocouples as identified in Figure 10-2. The parallel thermocouples on the specimen and tube or rod and tube should read the same temperature within about two degrees centigrade under steady state conditions although the temperature can vary axially somewhat from one end of the tube towards the other end. This axial variation should be kept as small as possible. The primary consideration is that the vitreous silica tube and rod must be at the same temperature along their length so that the expansion of the tube is compensated by the expansion of the rod resulting in zero change at the measuring transducer except for that length of tube which holds the specimen. It is for this length that the vitreous silica correction is needed because the specimen is taking the place of the rod. It is therefore necessary to get the tube and rod to the same temperature when the thermal expansion reading is made. A uniform temperature distribution is attained by placing external resistances between the taps on the outside of the multi-tap furnace, between which the internal temperature is too high; and this correction is determined

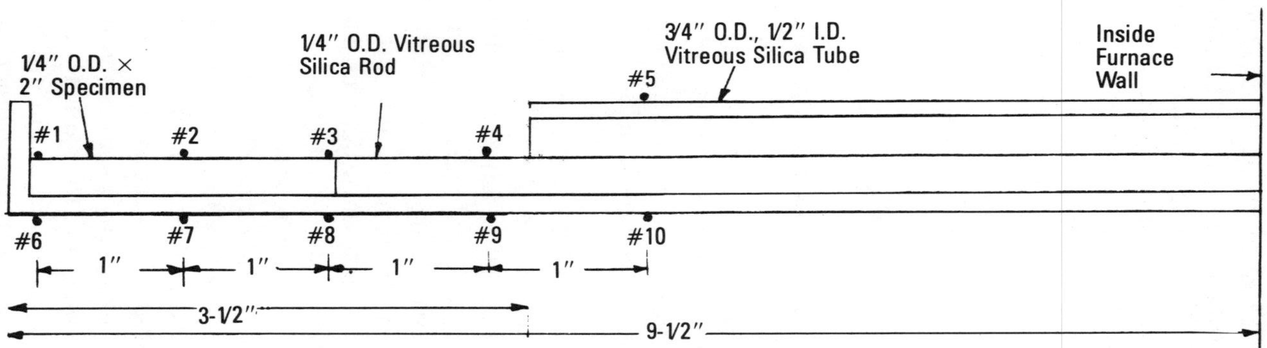

STEADY STATE TEMPERATURE IN °C AT THERMOCOUPLES

#1	#2	#3	#4	#5	#6	#7	#8	#9	#10
452	452	452	444	430	452	452	450	448	430
735	737	738	728	706	734	735	735	727	711
1036	1037	1037	1023	1000	1034	1039	1043	1024	1000
1060	1060	1057	1049	1028	1056	1056	1061	1050	1022
1162	1161	1162	1152	1126	1159	1158	1165	1152	1122

FIGURE 10–2.—Typical temperature distribution over specimen, vitreous silica tube and vitreous silica rod.

under steady state conditions. The external taps are electrically wired to the ends of each zone of the furnace. Ultimately, when the specimen is tested, it will be heated at a uniform rate, and it is assumed that the steady state temperature distribution holds for the uniform rate of heating condition. From experience, it has been found that a good steady state temperature distribution will be satisfactory for the uniform heating rate tests. Any deviations or errors will be small and will be included in the correction when the system is calibrated with a secondary standard.

In Figure 10-2 thermocouples were placed at one-inch intervals over a four-inch length measured from the end of the tube because in this example it had been shown from previous measurements, not given here, that the remainder of the tube-rod assembly was being heated uniformly. The data obtained in Figure 10-2 were a check in the immediate specimen area only. With a new furnace, thermocouples would be placed over the entire length at one-inch intervals. The center of the specimen is located at the center of the furnace.

The data in Figure 10-2 show that the temperature over the specimen length varies by a maximum of 3°C at all the temperature levels investigated. In general, the temperature variation is within about 5°C at all thermocouples with a few exceptions. This is considered a very good temperature distribution; typically the temperature distribution is poorer than that depicted in Figure 10-2. In general, a minimum variation between parallel thermocouples is desirable because the correction determined with the secondary standard will be minimized.

Transducer and Recorder

The transducer used to measure the thermal expansion must be calibrated for deflection as a function of its voltage output or, more particularly deflection, as a function of chart divisions the pen moves on the X–Y recorder used for the measurement. The pen movement is directly proportional to the voltage output of the transducer. A typical recorder is a Moseley Autograf Model 7001A made by Hewlett Packard. These X–Y recorders are generally used since the thermal expansion can be recorded on one axis and the temperature on the other axis. The detail of calibrating each dilatometer has been presented in the chapter covering each dilatometer. For the immediate purpose it is sufficient to say that the transducer must be calibrated in order to obtain accurate results.

The X–Y recorder must also be calibrated before each test. This is done electrically using a Model 8662 precision L&N potentiometer or an equivalent. The millivolt output of the potentiometer is fed to the X and Y-axis of the X–Y recorder and the deflection of the pen noted. If, for example, the X–Y recorder is set for 15 mV full scale, the pen should deflect full scale (10 inches on the Model 7001A X–Y recorder) when 15 mV is set on the potentiometer. The full scale or range will be determined by the expected thermal expansion, which will be explained later. The instructions for the use of the potentiometer for these calibrations are given in the instruction manual for the potentiometer. In the case of the Model 8662 potentiometer, the standard cell normally set at room temperature for the thermocouple being used is set at zero, and the main dial is set at the proper millivolt setting for which the calibration is desired. The output of the potentiometer is the difference between the dial setting and the standard cell setting. When the standard cell is set at zero, the potentiometer output is that set on the dial. If the standard cell is set at some value, say 1.00 mV, this value will have to be subtracted from all main dial settings to obtain the correct potentiometer output.

For the temperature axis, the test temperature range and the thermocouple output will determine the recorder range that will be used and calibrated.

During a test, it is advisable to check the temperature indicated by the X–Y recorder with the potentiometer. The measuring thermocouple on the specimen should be wired to a 2-pole knife switch so that the technician can switch from the recorder to the potentiometer rapidly. Often electrical interference will affect the X–Y recorder or the thermocouple resulting in an erroneous temperature measurement.

Calibration Procedure

Now that the furnace temperature distribution has been optimized and the transducer and X–Y recorder calibrated, the thermal expansion of chemically pure platinum, a secondary standard, can be measured to obtain a system calibration. This will permit correction for the inaccuracies in our individual calibrations and any other system eccentricities that would otherwise result in errors, such as thermal expansion variations between the vitreous silica tube and rod.

A typical thermal expansion calibration curve using pure platinum is shown in Figure 10-3. The temperature axis is labelled with a template corresponding to a chromel-alumel thermocouple output as shown in Figure 10-4. Since the same thermocouple is used in routine tests to 1000°C, it was convenient to make a template from plexiglas so that the temperature axis could be easily and uniformly identified every 100°C from test to test. Similar templates can be cut for other thermocouples. The laboratory in which these tests are made is temperature and humidity controlled so that the starting temperature is always 25°C.

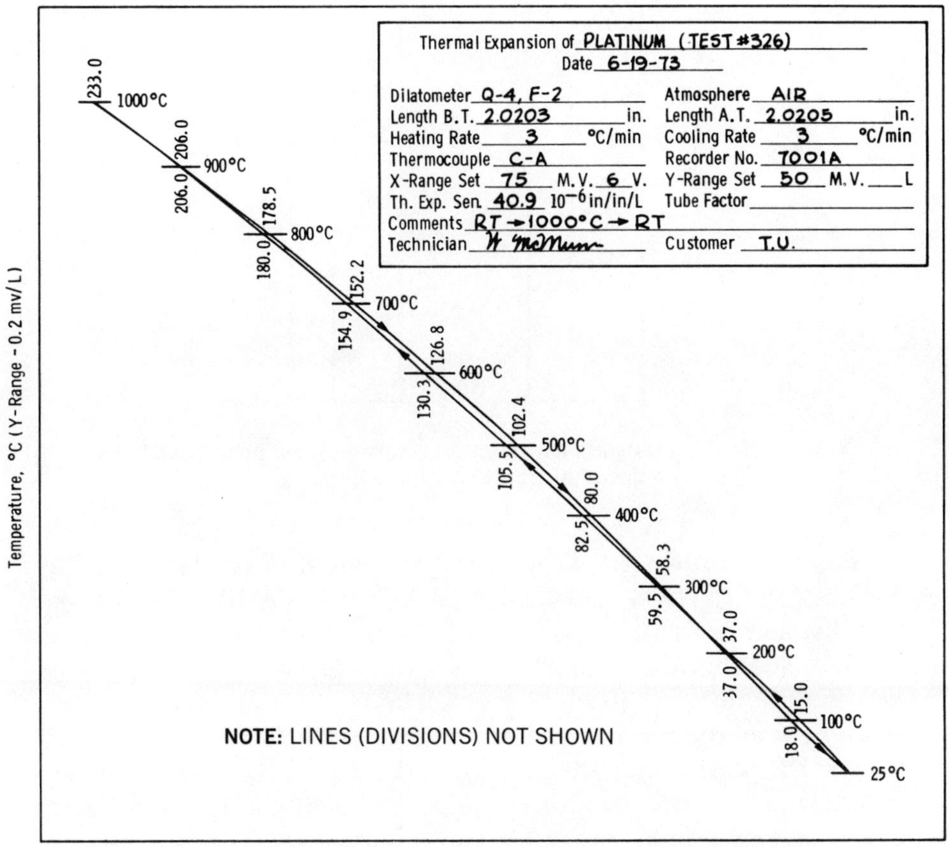

Thermal Expansion (X-Range, 40.9 × 10^{-6} in/inL)

FIGURE 10-3.—Typical *X–Y* recorder curve of temperature versus linear expansion.

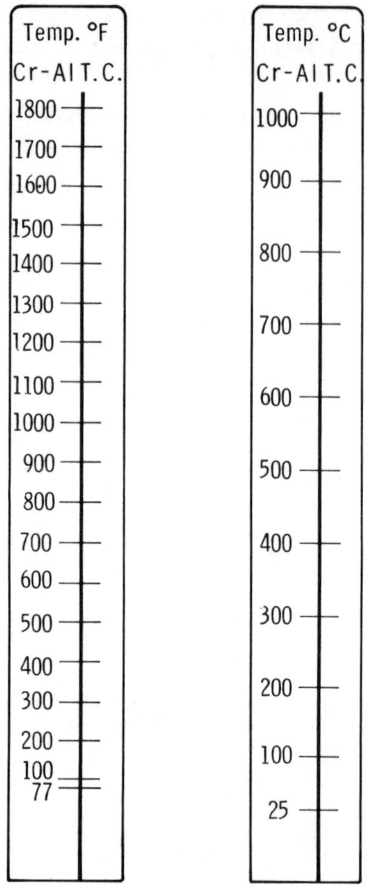

FIGURE 10–4.—Plexiglass templates for chromel-alumel thermocouple for temperature axis of *X–Y* recorder curve.

From the calibration of the transducer, on the 75 mV range of the X-axis of the Hewlett-Packard 7001A X–Y recorder, it is known that one division is equal to 40.9×10^{-6} inches/inch (m/m) expansion. Actually the thermal expansion of a 2-inch specimen was measured but the calibration coefficient is given in inches/inch. At each 100°C increment on the heating and cooling cycles, the divisions of expansion on the papergraph are counted with 25°C as the zero expansion point. Other pertinent information is given in the legend of each graph as shown in Figure 10-3. The number of divisions expansion at each temperature is tabulated as shown in columns (1) and (2) in Table 10-1. The number of divisions expansion at each temperature

Table 10–1
Calibration of Dilatometer Containing Q-4 Vitreous Silica Tube and Rod Using Chemically Pure Platinum Standard (Test 326)

MEASURED THERMAL EXPANSION				
(1) Temperature, °C	(2) Lines deflection on Y-axis	(3) 10^{-6} inches/inch at rate 40.9×10^{-6} inches/inch division on Y-axis	(4) ASTM thermal expansion of pure platinum in 10^{-6} inches/inch	(5) Correction factor for dilatometer in 10^{-6} inches/inch
25	0	0	0	0
100	15.0	614	675	61
200	37.0	1513	1604	91
300	59.5	2434	2558	124
400	82.5	3374	3537	163
500	105.5	4315	4541	226
600	130.3	5329	5572	243
700	154.9	6335	6629	294
800	180.0	7362	7713	351
900	206.0	8425	8823	398
1000	233.0	9530	9961	431
900	206.0	8425	8823	398
800	178.5	7301	7713	412
700	152.2	6225	6629	404
600	126.8	5186	5572	386
500	102.4	4188	4541	353
400	80.0	3272	3537	265
300	59.3	2384	2558	174
200	37.0	1513	1604	91
100	15.0	614	675	61
25	0	0	0	0

level is multiplied by the coefficient 40.9×10^{-6} inches/inch div. to yield the total uncorrected thermal expansion from 25°C to each temperature in 10^{-6} inches/inch as given in column (3). The values in column (3) are the expansion of platinum from 25°C to the respective temperatures in 10^{-6} inches/inch of specimen without vitreous silica corrections or any other corrections that would introduce errors. Column (4) lists the true thermal expansion of platinum as reported in the ASTM standards. The values in column (3) are compared to the values in column (4) and their difference is given in column (5), which indicates the corrections for the system at each temperature. Another way of looking at this is to say: what must be added to the values in column (3) to make them the same as column (4) are the values given in column (5). The corrections, column (5), are next plotted as a function of temperature, column (1) and are shown in Figure 10-5 as Test 326. Since it is known from experience that the correction curve as described to this point is variable, additional calibration tests must be conducted to determine the magnitude of the variability and to determine an average correction for the system. The results of two such additional tests are presented in Figure 10-5 along with the average of the three tests which was used as the correction curve for the system.

New Vitreous Silica Tube and Rod Combination

As is obvious from the data in Figure 10-5, the corrections on the heating and cooling cycles are different so that different corrections must be used for specimens tested on heating and cooling. The maximum variability occurs at 1000°C and is about 100 × 10^{-6} inches/inch. by using an average correction one reduces this variability to about ±50 × 10^{-6} inches/inch; and if a material having a total thermal expansion of 10,000 × 10^{-6} inches/inch were tested, the error would amount to 50 parts in 10,000 or 0.5%, which is quite low. The correction in Figure 10-5 is for a particular vitreous silica

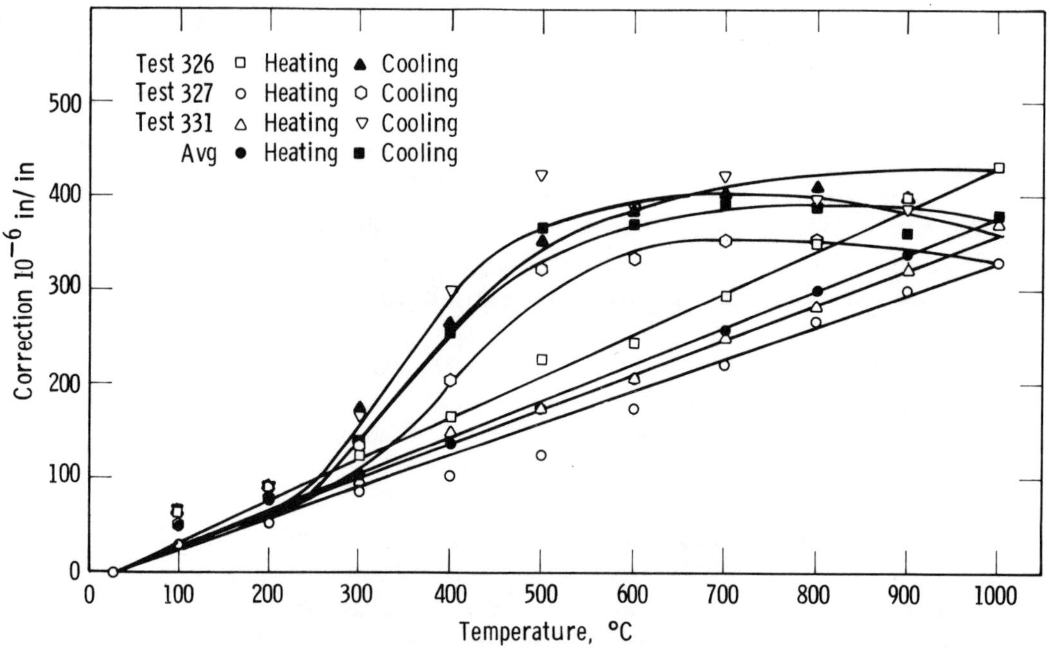

FIGURE 10–5.—Calibration of dilatometer containing Q-4 fused silica tube and rod using chemically pure platinum standard (short term).

tube vitreous silica rod combination (Q4) and only for this combination. These corrections were obtained from tests conducted during a short term of three consecutive days and give no information as to the long-term change in corrections when the tube and rod combination is used daily in tests to 1000°C on glasses, metals, and other materials that may contaminate the vitreous silica changing its thermal expansion properties. Devitrification starts at 1000°C without contamination and is a factor. Oil from the technician's hands also introduces contamination, which accelerates devitrification. It has been stated by vitreous silica manufacturers that vitreous silica should be handled with white gloves to minimize the effect of contamination from the operator's hands. The use of white gloves by the technician is not a practical procedure so that frequent calibrations are needed to monitor the changes in the thermal expansion of the vitreous silica tube and rod combination.

Effects of Repeatedly Cycling Vitreous Silica to High Temperature

During the course of daily thermal expansion testing to temperatures as high as 1000°C with the same vitreous silica tube and rod, it was interesting to note how the vitreous silica expansion and hence the dilatometer corrections changed with time. Records of the corrections were kept, and Figure 10-6 shows the correction curves obtained using the same tube and rod combination with the same dilatometer over a period of approximately seven months. As shown in Figure 10-6, the average correction appears to have increased progressively from the test conducted 6-27-73 to the test conducted 4-17-74 with the exception of the test conducted on 4-14-74. During the entire period from 6-27-73 to 4-17-74, the tube and the rod were used daily in tests on glass to about 700°C maximum, on plastics to 150°C maximum, and on metals and ceramics to about 1000°C. Other vitreous silica tube-rod combinations not reported here were investigated but the same trend in increasing corrections was always present and broken by a test during the period of investigation. It seems hardly possible that the thermal expansion of the vitreous silica tube and rod is not stable and increases or decreases with usage. Perhaps the tube and the rod were used to higher temperatures and under more contaminating conditions during the period which the corrections reverse in trend. The rate of heating the

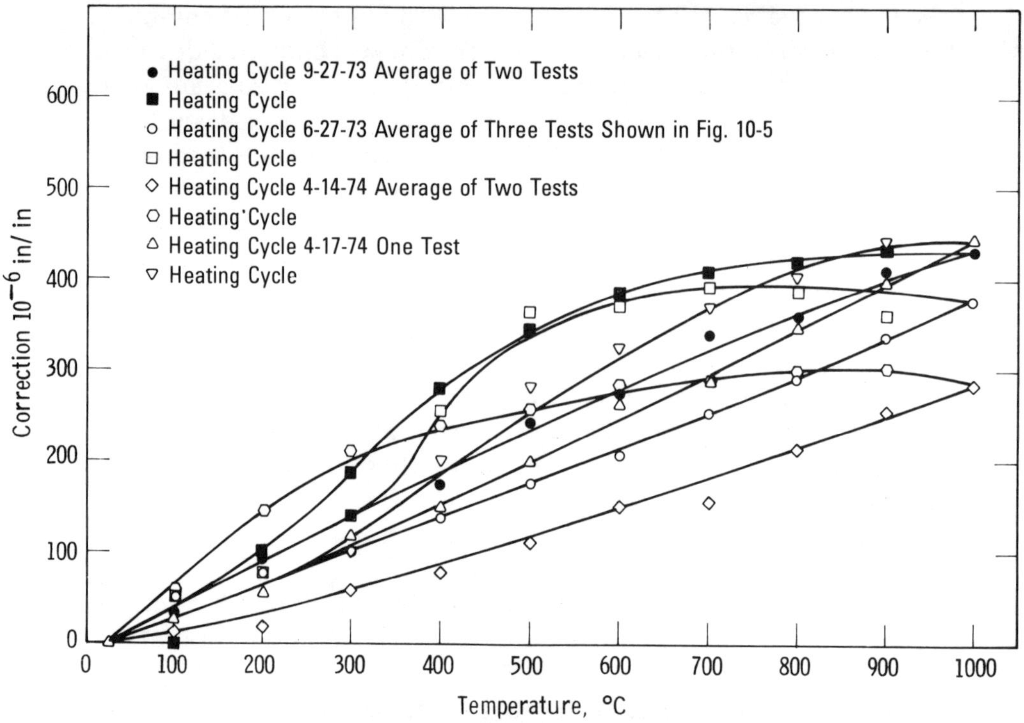

FIGURE 10–6.—Calibration of dilatometer containing Q-4 fused silica tube and rod using chemically pure platinum standard (long term).

platinum specimen could be the cause of the reversal. Every effort, however, is made to conduct all tests at the same heating rate using a program controller. Since experience has shown that the vitreous silica corrections can change as discussed above, frequent calibration of the dilatometer is required for the best accuracy. If one is going to conduct a very critical thermal expansion test on a material, it is best to conduct a calibration test before and after the specimen is tested in order to obtain the most accurate data. It would also be wise to run each test several times to reduce variability.

Average Vitreous Silica Corrections versus Calibration Corrections

The average accepted coefficient of thermal expansion of vitreous silica from 25°C to 1000°C is 0.5×10^{-6} inches/inch °C. If corrections based on this coefficient are compared to the data shown in Figure 10-6, it is evident that in general all corrections on the heating cycle lie below the average vitreous silica corrections. It must be remembered that the data in Figure 10-6 in most cases are the average obtained from several tests conducted around those dates so that the individual tests that were used to obtain these averages deviated more than the curves in Figure 10-6 indicate.

On the cooling cycle the data in Figure 10-6 show that the measured corrections are, in general, higher than the accepted values of vitreous silica. This is not the fault of the vitreous silica; it expands and contracts at the same rate. It is probably the result of the temperature gradient over the vitreous silica tube and rod system. The temperature distribution, as explained in the section on the furnace, is set under static conditions where the temperature is allowed to stabilize before a reading is taken. The data obtained in Figure 10-6 were obtained while the specimen was heated and cooled at 3°C/min., and the resultant temperature gradients may be the reason the cooling cycle data is higher than the generally accepted vitreous silica thermal expansion values. The difference cited is not serious as long as the corrections are repeatable from test to test. One can use the average accepted values and still obtain data that will solve problems. If highly accurate data are required, calibrations should be conducted before and after the specimens are tested.

High Expansion Standard versus Low Expansion Standards

As one goes about his daily task of measuring the thermal expansion of solids and calibrating the apparatus, a question arises. What should be the thermal expansion of a secondary standard suitable for calibrating equipment used in daily thermal expansion testing? The ASTM standard E228 requires the use of chemically pure platinum from −195°C to 1000°C or OFHC copper from −195°C to 100°C. The average thermal expansion coefficient of platinum is 10.19×10^{-6} inches/inch in the range 0°C to 1000°C and 8.00×10^{-6} inches/inch in the range −195°C to 0°C and copper 16.99×10^{-6} inches/inch from 0°C to 100°C and 13.83×10^{-6} inches/inch from −195°C to 0°C. These are high coefficients of thermal expansion and these secondary standards are suitable for calibrating dilatometers used for testing materials with similar thermal expansions. However, if one is testing low expanding glasses or vitreous silica, platinum and copper are not the best secondary standards to use in calibrating the dilatometer. Experience has shown the author that the calibration values obtained with high expanding secondary standards are not suitable to correct the thermal expansion data measured with low expanding materials such as vitreous silica. If one were to use the correction data obtained with platinum and apply it to data obtained by testing vitreous silica, one would find his results could be in error by as much as a factor of two. One just cannot use corrections generated with platinum or copper on data obtained with low expanding materials like vitreous silica because the thermal expansion of the two materials is radically different. One must use a standard that has about the same expansion as the material he is going to test in order to obtain the optimum accuracy with the dilatometer. The platinum has a higher heat capacity and better thermal conductivity than vitreous silica in addition to the radical difference in thermal expansion that contributes to the error. But the biggest error comes from the dilatometer itself. No dilatometer will give perfect results; each has an inherent lack of accuracy, and this inaccuracy is a larger percentage of the thermal expansion when low expanding materials are being tested.

For years it has been accepted practice to purchase a secondary standard from the U.S. Bureau of Standards with an expansion coefficient of about 10×10^{-6} inches/inch °C and to use it for calibrating dilatometers which are used to test materials with higher, equivalent, and lower coefficients. In recent years it has been realized that this practice is not good. For this reason the Bureau of Standards has adopted the policy of supplying secondary standards with high, low, and intermediate coefficients of thermal expansion so that the experimenter can choose a secondary standard that has a thermal expansion coefficient near the material he is testing. If a laboratory is testing all kinds of materials having high, intermediate, and low thermal expansion coefficients, it should have a corresponding variety of secondary standards and separate correction data for the dilatometer using each standard.

Heating Rate Effects

The rate at which a specimen is heated should be slow enough to insure that the specimen, the rod, and the tube are at the same temperature at any instant. Enough time must be allowed for heat transfer to take place from the heating elements of the furnace to the tube, the rod, and the specimen. The heat transfer will take place by conduction, convection, and radiation. If the heating rate is too slow, high accuracy may be maintained but it will merely take more time to run the test, and the testing cost will increase. If, on the other hand, the heating rate is too fast, inaccuracies may result and will be somewhat proportional to the heating rate. When the heating rate is too fast, the tube will expand faster than the rod and specimen; and in effect the tube will be longer, hiding or reducing the thermal expansion.

Over the years heating rates of the order of 1°C every three minutes have been advocated but these are too slow, resulting in a time consuming, expensive test which is frowned upon by industry. It has been the author's experience that a heating rate of 3°C/min. for a vitreous silica tube dilatometer is satisfactory. With this heating rate a technician can set up a specimen in the morning, measure its thermal expansion to 1000°C, and cool it near the starting temperature in one shift. The specimen does not cool to room temperature by the end of the shift because of the mass of the resistance-wound furnace, which holds the heat; but the cooling curve is close enough to the starting temperature (room temperature) to be very accurately extrapolated to room temperature.

For tests to 1600°C using the alumina tube dilatometer, the heating and the cooling of a specimen are accomplished over two shifts, or about 16 hours, which means that the specimen is heated and cooled at a rate of 3.5°C/min., an acceptable rate. However, it would be better to heat the specimen more slowly than 3.5°C/min. because of the large coefficient of thermal expansion and contraction of alumina. Again a high mass platinum resistance-wound furnace is used to heat the specimen, and the rates are reduced as room temperature is approached on the cooling cycle, requiring some extrapolation of the data.

For tests to 2200°C using the tantalum tube dilatometer described in Chapter 5, the heating and the cooling rates are about 4.5°C/min. because it is desirable to complete the test in two shifts. Since the tube and the rod are metals, good heat conduction takes place and a faster heating rate can be used. This dilatometer has a low mass and the heating rate can be controlled over the entire heating and cooling cycles.

Secondary Standards

The best way to calibrate a dilatometer is to use a primary standard. However, primary standards are so expensive that an alternate standard must be used. The alternate standard is a secondary standard, which is comparatively inexpensive. Secondary standards are supplied by the U.S. Bureau of Standards. As of this writing, the Bureau of Standards can supply the standard reference materials (SRM) listed in Table 10-2, which are also given in reference 1.

The secondary standards are prepared by testing a number of specimens from a given heat of material and concluding that the thermal expansion of the tested specimens (primary standards) is representative of the whole heat. The individual specimens for sale are then cut from the bar stock and the thermal expansion measured with the primary piece assigned to them. Extreme care is exercised in testing the material. Basically the thermal expansion is measured optically and the specimen is heated with a resistance-wound furnace. The method of holding the specimen at a predetermined temperature until thermal equilibrium is established is used. A continuous curve as advocated by the author is not used. The method used by the Bureau of Standards is very accurate, and considerable effort is expended to attain thermal equilibrium. Such effort is too expen-

Table 10–2
Standard Reference Materials

These SRM's cover the temperature range from 20 K to 1900 K having coefficients of thermal expansion over the range of 0.5 to 25 × 10^{-6}/K

SRM	Type	Temperature range, K	Diameter, mm	Length, mm
731-L1	Borosilicate Glass	80–680	6.4	51
731-L2	Borosilicate Glass	80–680	6.4	102
731-L3	Borosilicate Glass	80–680	6.4	152
732-L1	Sapphire	IN PREP		
732-L2	Sapphire	IN PREP		
732-L3	Sapphire	IN PREP		
736-L1	Copper	20–800	6.4	51
736-L2	Copper	20–800	6.4	102
736-L3	Copper	20–800	6.4	152
737-L1	Tungsten	IN PREP		
737-L2	Tungsten	IN PREP		
737-L3	Tungsten	IN PREP		
739-L1	Fused Silica	80–1000	6.4	51
739-L2	Fused Silica	80–1000	6.4	102
739-L3	Fused Silica	80–1000	6.4	152

U. S. Department of Commerce
Maurice H. Stans
Secretary

National Bureau of Standards
L. M. Branscomb, Director

Certificate of Analysis

Standard Reference Material 739

Fused-Silica Thermal Expansion

R. K. Kirby and T. A. Hahn

Thermal Expansion as a Function of Temperature

T	Expansion $\Delta L/L_{293}$	Expansivity α	T	Expansion $\Delta L/L_{293}$	Expansivity α
80 K	-1×10^{-6}	-0.70×10^{-6}/K	320 K	$13._5 \times 10^{-6}$	$+0.53 \times 10^{-6}$/K
90	$-7._5$	−0.61	340	$24._5$	0.56
100	−13	−0.53	360	36	0.58
110	−18	−0.46	380	$47._5$	0.60
120	$-22._5$	−0.38	400	$59._5$	0.61
130	−26	−0.31	420	72	0.62
140	$-28._5$	−0.24	440	85	0.63
150	$-30._5$	−0.17	460	97	0.63
160	−32	−0.10	480	110	0.63
170	$-32._5$	−0.04	500	122	0.63
180	$-32._5$	+0.02	520	135	0.62
190	−32	0.08	560	159	0.61
200	−31	0.13	600	183	0.59
210	$-29._5$	0.19	640	206	0.56
220	$-27._5$	0.23	680	228	0.54
230	−25	0.28	720	249	0.51
240	−22	0.32	760	269	0.49
250	$-18._5$	0.36	800	288	0.47
260	$-14._5$	0.39	840	307	0.44
273	− 9	0.43	880	324	0.42
280	− 6	0.45	920	340	0.40
293	0	0.48	960	356	0.38
298	$+2._5$	0.49	1000	371	0.37

This SRM is available as a rod 6.4 mm (¼ inch) in diameter; L1 is 51 mm (2 inches) long, L2 is 102 mm (4 inches) long, and L3 is 152 mm (6 inches) long. (Note: Inquiries for longer continuous rods may be directed to the Office of Standard Reference Materials, National Bureau of Standards.) The rods that make up this SRM were obtained as a single drawing from an ingot of fused silica, at least 99.8 wt percent pure. These rods have been annealed by soaking at 1373 K for 7 hours and cooling to 1173 K at 12 K per hour.

The above values of expansion and expansivity were calculated from a least squares fit to expansivity measurements made on five specimens. Linear interpolation can be used between tabulated values without introducing a significant error. The expansion of this material may be altered, however, if the material is heated to temperatures above 1020 K. A description of the experimental method, fitting procedure, and estimate of uncertainties is given on the back of this certificate.

The overall coordination and evaluation of data leading to certification of this SRM was performed by R. K. Kirby and T. A. Hahn.

The technical and support aspects involved in the preparation, certification, and issuance of this Standard Reference Material were coordinated through the Office of Standard Reference Materials by R. E. Michaelis.

Washington, D. C. 20234
May 12, 1971

J. Paul Cali, Chief
Office of Standard Reference Materials

FIGURE 10–7—A typical certificate of analysis for a standard reference material from the U.S. Bureau of Standards

sive for industry, however. For industry it is best to use the secondary standards supplied by the Bureau of Standards and employ the technique of obtaining a continuous curve of temperature vs. thermal expansion. The Bureau of Standards data are probably accurate to ±0.5%. This order of accuracy is always desirable but not practical because of cost. Results obtained by the continuous recording of thermal expansion and temperature can be as good as ±2% on a routine daily basis, and this degree of accuracy is more than enough to solve most problems.

One could generate his own secondary standards with the same technique used by the Bureau of Standards. This is not advisable, however, because the Bureau of Standards can do the job cheaper and better. Also, if all laboratories use the same standards, as those from the Bureau of Standards, there will be a greater uniformity of data generated among laboratories. The author recommends that the secondary standards supplied by the Bureau of Standards be used to calibrate all dilatometers.

None of the secondary standards available to date are suitable for calibrations to 1600°C in air. For these ranges it is best to use chemically pure platinum, which is available from a number of suppliers. To date there are not enough secondary standards to cover very many temperature ranges. The best one can do for the present is to purchase all the secondary standards available and use the one closest to the thermal expansion of the material being tested.

A typical certificate of analysis for a fused silica secondary standard is given in Figure 10-7. From this the reader can see how the Bureau of Standards presents its calibration standards data. The overall coordination and evaluation of data leading to certification was done by R. K. Kirby and T. A. Hahn.

Reference

1. NBS Special Publication 260, "Standard Reference Materials Price and Availability List." U.S. Department of Commerce, National Bureau of Standards.

References Not Cited

1. J. B. Conway, "Properties of Some Refractory Metals," General Electric Atomic Products Division, GEMP-375, June 29, 1965.

2. J. L. White, "The Suitability of Platinum, Molybdenum, Tantalum and Tungsten as High Temperature Length Standards," NRL Report 5159, Naval Research Lab., Washington, D.C., June 17, 1958.
3. I. E. Campbell, *High Temperature Technology*, John Wiley and Sons, New York, 1956.
4. R. R. Matheson, "G. E.'s Lamp Research Leads to Unique Ceramic Material," *Ceramic Age*, June, 1963.
5. R. E. Taylor and G. L. Denman, Thermal Expansion, 1973, *American Institute of Physics Conference Proceedings No. 17*, American Institute of Physics, New York, 1974.
6. M. G. Graham and H. E. Hagy, Thermal Expansion, 1971, *American Institute of Physics Conference Proceedings No. 3*, American Institute of Physics, New York, 1972.
7. Symposium on the Thermal Expansion of Solids, *Journal of Applied Physics*, 41, 13, 1970, pp. 5043-5154.
8. J. B. Austin, "A Vacuum Apparatus for Measuring Thermal Expansion at Elevated Temperatures with Measurements on Platinum, Gold, Magnesium, and Zinc," Physics, 3, 1932, pp. 240-267.
9. F. C. Nix and D. McNair, *Physical Review*, 61, 1942, pp. 74-78.
10. F. C. Nix and D. McNair, *Physical Review*, 60, 1941, p. 597.
11. T. H. Rubin, W. Altman and H. L. Johnston, *Journal of the American Chemical Society*, Volume 76, 1954, p. 5289.
12. T. A. Hahn, "Thermal Expansion of Copper from 20 to 800 K," *Journal of Applied Physics*, 41, 13, 1970.
13. R. C. Rossi, "Thermal Behavior of Ceramic Composites," *Air Force Report No. SAMSO-TR-69-176*, Customer Services, Clearinghouse, U.S. Dept. of Commerce, Springfield, Va. 22151.
14. D. J. Michel, "Bibliography on Thermal Expansion in Intermetallic Compounds," *Atomic Energy Commission Report No. NYO-3560-17*, Clearinghouse, U.S. Dept. of Commerce, Springfield, Va. 22151.
15. N. S. Rasor, "Thermal Properties of Materials," *Wright Air Development* Center, WADC-TR-56-400 Part 1, ASTIA Document No. AD 118144, 1957.
16. R. H. Keith, "A Kinematic Dilatometer," Technical Report 65-107, The University of Dayton Research Institute, Dayton, Ohio 45409.
17. C. P. Kempter, *Journal of Chemical Physics*, 43, 5, 1965, pp. 1739-1742.
18. G. K. White, "Measurement of Thermal Expansion of Low Temperatures," *Cryogenics*, March, 1961.
19. S. F. Jacobs, "Ultraprecise Measurement of Thermal Coefficients of Expansion," *Applied Optics*, 9, 11, 1970.
20. R. A. Schapery, "Thermal Expansion Coefficients of Composite Materials Based on Energy Principles," *Journal of Composite Materials*, 2, 3, 1968.
21. W. Souder, "Measurements on the Thermal Expansion of Fused Silica," Scientific Papers of the U.S. Bureau of Standards, Vol. 21, S524.
22. L. T. Loyd, "Recording Quartz Differential Dilatometer," Argonne National Lab—S372, 1959.
23. S. C. Fulkeson, "Apparatus for Determining Linear Thermal Expansion of Materials in Vacuum or Controlled Atmosphere," Oak Ridge National Lab—2856, 1960.
24. P. S. Gaal, "Quartz Pedestal Dilatometer," Temperature Measurement Society, Sixth Conference and Exhibit, April 1969.

CHAPTER 11

Dilatometers for Measurements from −271°C to 25°C

Introduction

When it is desirable to measure the thermal expansion of solids from room temperature to near absolute zero or about −271°C (2°K), one must use liquid helium as the cooling medium. Although the problem of liquid helium storage, its transfer to the dilatometer from the storage tank, and suitable control equipment have been developed and used, it is still comparatively difficult to conduct tests with a liquid helium dilatometer. Precautions which are not normally considered must therefore be practiced.

The need for a liquid helium dilatometer arose from the use of materials in space flight where low temperatures are encountered. The development of the superconducting generator and nuclear fusion also require materials to withstand low temperatures so that the liquid helium dilatometer is necessary to measure the thermal expansion of solids for these applications.

Six cryogenic dilatometers are described in this chapter. These include four cryogenic dilatometers designed by other investigators, one proposed by the author, and a commercially available cryogenic system that is suitable for a cryogenic dilatometer. It must be emphasized that the author's cryogenic dilatometer is a proposed one that has not been proven in the laboratory. However, it is the author's opinion that it will perform satisfactorily with some minor modifications that will become apparent during testing.

Author's (Valentich) Proposed Cryogenic Dilatometer

The author's proposed cryogenic dilatometer, which will be identified as the Valentich Cryogenic Dilatometer, is described in this section. The cryostat is similar to one successfully used by the Westinghouse Research Center for tensile testing to about −268°C (5°K). Based on this experience, it is the author's belief that this cryostat will perform satisfactorily for the cryogenic dilatometer.

Description of Cryostat

The cryostat, shown in Figure 11-1, is constructed from AISI 304 stainless steel sheets 0.018 to 0.025 inch thick. The wall thickness of each chamber must be kept small to help minimize heat leaks and initial heat content. Styrofoam is used for insulation as indicated in Figure 11-1. The four concentric cylinders: items 1, 2, 3, and 4 in Figure 11-1, are heliarc welded to stainless steel flanges at the top and are sealed from one another with o-rings. The flanges are held together with eight screws: item 20, around the cryostat. The lid of the cryostat is a Styrofoam insulated liquid nitrogen chamber welded to a flange which is locked to the flanges of items 1 to 4, with eight hinged screws: item 18. An o-ring forms the seal between the lid flange and the adjacent flange. The bottom plates of the chambers are heliarc welded to the cylinders. The cryostat is about 5½ inches in diameter by 14½ inches long, excluding the tubes protruding out of the lid. The Styrofoam chambers and liquid nitrogen chamber are about ½ inch wide radially and at the bottom. The inner chamber has a 2½ inch ID by 6 inches long. The wall of the copper susceptor is about ⅛ inch thick; and an insulated heater is attached to it.

Two hoops: items 24 and 25, are spot welded to the outer

1-7	304 SS Cylinders
8	Liquid helium inlet
9	Sealing adapter
10	Copper susceptor
11	Heater
12	Tube inlet
13-14	Pressure exhaust and liquid nitrogen inlet (4 items)
15	Control thermocouple and heater power inlet
16	Liquid helium exhaust
17	3 Styrofoam chamber vents
18	8 Locking hinge screws and nuts
19	Styrofoam insulation
20	8 Cap screws
21-23	O-rings
24-25	Upper and lower mounting plates and hoops
26	6 304 SS Hoops
27	6-ringe sealing adapter
28	Capillary tube

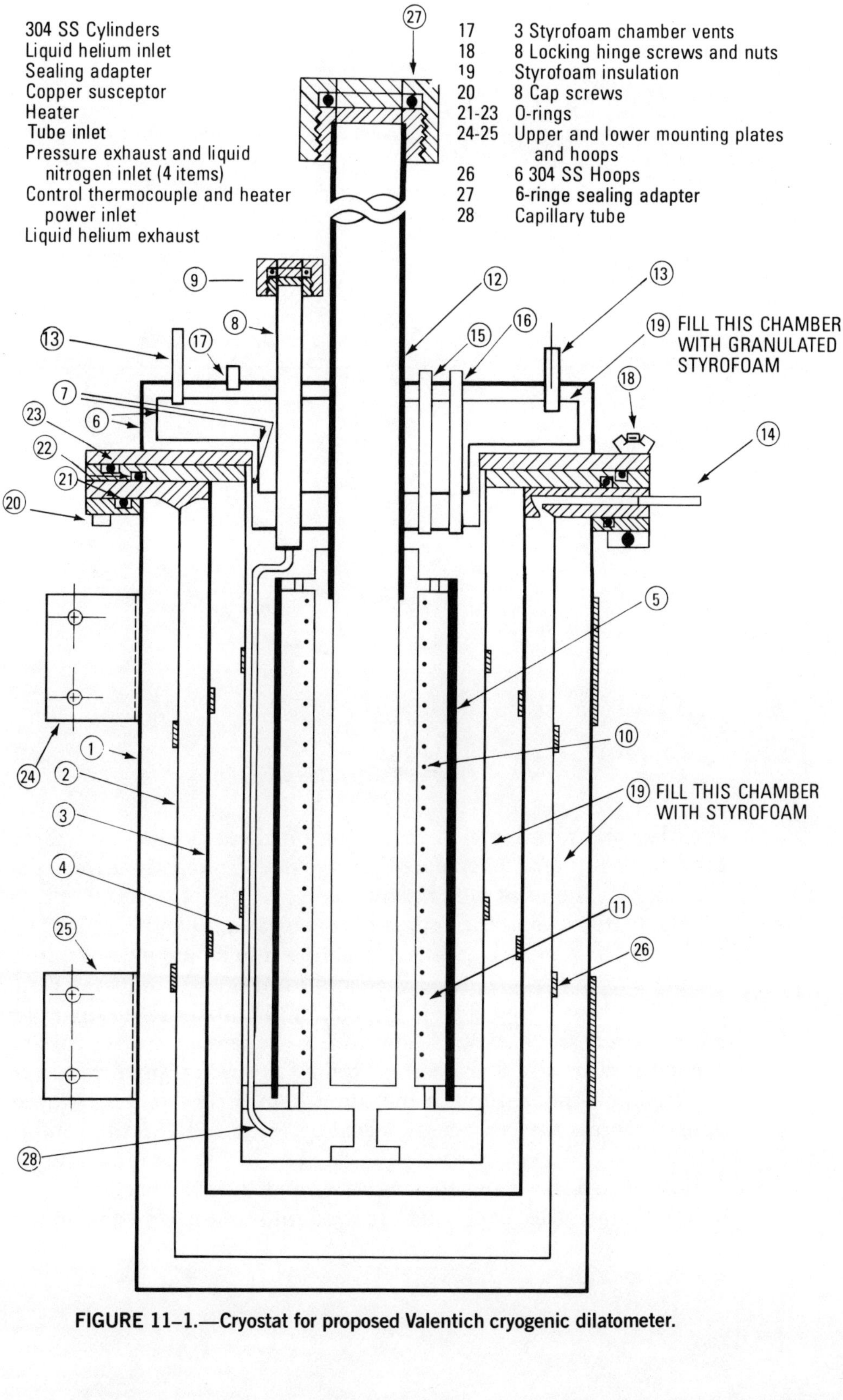

FIGURE 11–1.—Cryostat for proposed Valentich cryogenic dilatometer.

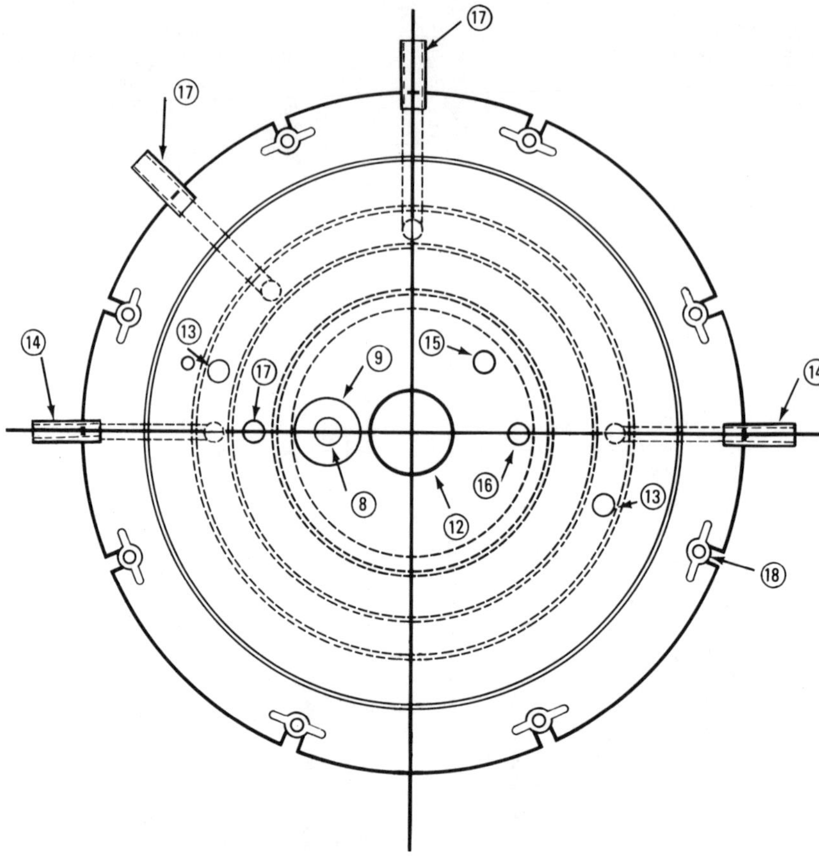

FIGURE 11–1.—Top View.

chamber and terminate as mounting brackets as shown in Figure 11-1. Six additional hoops are spot welded around the three inner chambers to increase their rigidity.

The baffle around the copper susceptor is a 1/16 inch-thick 304 SS shell, which serves as a thermal shield. An inlet and exhaust: items 13 and 14, are provided for the liquid nitrogen chambers. Vents: item 17, are provided for all Styrofoam chambers to dissipate any pressure built up that may occur during the temperature excursions experienced by the Styrofoam. Liquid helium is transferred to the specimen chamber through the tube: item 8. The sealing adapter: item 27, forms the compression seal between a 304SS tube and the cryostat.

The electrical wiring to the heater and the thermocouple wires for the temperature controller are brought into the cryostat through the tube: item 15. A suitable sealing element, like a Conax seal, can be used to seal around the wires. A pressure gage is attached to the tube: item 16, to control the venting of the liquid helium chamber.

Holes are provided at the bottom end of the copper susceptor to allow the liquid helium and its vapors to rise inside the shell: item 5, to cool the susceptor and hence the specimen. Similar holes are provided at the upper end of the susceptor to allow the vapors to escape to the atmosphere through the pressure valve.

Measuring Assembly

The measuring assembly is shown in Figure 11-2A. The specimen is 2 inches long by ¼ inch in diameter and is placed at the bottom of the closed vitreous silica tube. A thermocouple is wired to the specimen at midlength. A vitreous silica rod contacts the top of the specimen and transmits the specimen length change to the LVDT contacting the other end of the rod, as indicated in Figure 11-2A. The output of the LVDT is recorded on one axis of an X-Y recorder, and the thermocouple output on the other axis of the recorder, resulting in a continuous plot of expansion or contraction vs. temperature.

The LVDT is fixed to the open end of the vitreous silica tube using a split invar tube adapter which is attached to the vitreous silica tube with a small hose clamp. A lip at the bottom of the adapter prevents the hose clamp from falling off when it is loose. The vitreous silica tube, with the specimen and vitreous silica rod installed, is placed into a 304SS tube closed at the bottom end. A base plate is welded to the 304SS tube at the open end. In practice, the 304SS tube is permanently installed in the cryostat. An o-ring is used to form a compression seal between the cryostat and the outside diameter of the tube with the o-ring sealing adapter, item 27 in Figure 11-1.

A 304SS housing is placed over the measuring head; and an o-ring between the base plate and the housing flange, which is welded to the housing, forms a vacuum seal, as indicated in Figure 11-2A. Six cap screws hold the housing flange to the base plate. A vacuum port is provided in the base plate to evacuate the vitreous silica tube and the SS tube, which are back filled with helium gas to eliminate ice formation in the tube during testing. A pipe thread is provided in the base plate for attaching a Conax transducer gland to route the LVDT wiring, and the thermocouple attached to the specimen out to the recorder.

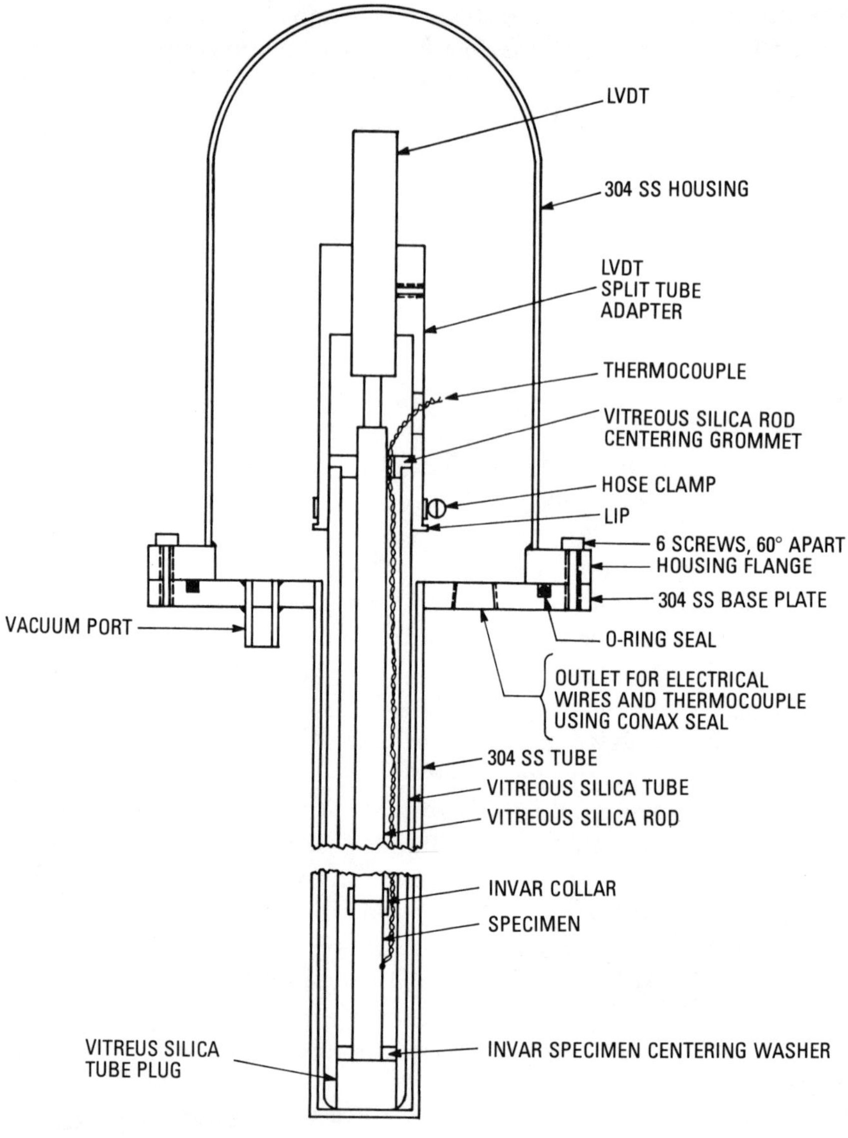

FIGURE 11–2A.—Measuring assembly for proposed Valentich cryogenic dilatometer.

Although not indicated in Figure 11-2A, the thermocouple is spot welded to the specimen and then wired to the specimen over an insulated section near the spot weld and along the vitreous silica rod at about two places. If bare wire thermocouples are used with two hole alumina insulators, the insulators should be wired to the specimen and vitreous silica rod which will result in a stable train that can be easily installed into the vitreous silica tube.

The specimen is centered at the bottom of the tube with an invar washer. The top of the specimen and bottom of the rod are held in line with a collar cemented to the rod as indicated in Figure 11-2B. At the open end of the tube an invar grommet is used to center the rod. This arrangement provides an annular space between the rod and the tube inside diameter to pass the thermocouple vertically out of the tube.

In an alternate method to center the internal elements of the vitreous silica tube, the vitreous silica rod can be centered by providing spherical guides near each end of the rod, as was done in the dilatometer designed by the European Organization for Nuclear Research, which is shown in Figure 11-4 of this chapter. A vertical slot would have to be cut through each sphere to permit the passage of the thermocouple. The specimen can be centered at the bottom of the tube with a washer and with the bottom of the rod using the collar which was described previously.

Operation of Dilatometer

The general procedure for operating the dilatometer is to purge the cryostat with helium gas and precool it with liquid nitrogen, which costs less than liquid helium. The specimen chamber is then purged with helium gas to blow out any remaining liquid nitrogen and then the liquid helium is introduced to cool the specimen to about −269°C (4.2°K). Liquid helium is transferred from the storage dewar by applying pressure into the dewar with a tank of helium gas. The rate of flow to the specimen chamber depends upon the size of capillary in the cryostat and is determined experimentally. After the specimen is cooled to about 4.2°K, an appropriate flow of liquid helium is continued while the heater is activated and controlled with a temperature controller. The specimen is heated at a rate which is determined experimentally based on the accuracy desired. Generally, however, a slower rate will result in more accurate data; economics will also be a consideration. A typical rate is 3°C (3°K) per minute. Consideration will also have to be given to determine the venting pressure from the test chamber. If the pressure in the chamber is too high, it may prevent the transfer of liquid helium from the supply dewar so that some optimum condition will have to be determined experimentally.

The specimen should be set up in the vitreous silica tube at room temperature, and put into the cryostat wired to an *X–Y* recorder. It is then evacuated and back filled with helium gas to prevent freezing. The cryostat is then precooled using liquid nitrogen and a measurement of specimen contraction made with the *X–Y* recorder at the lowest temperature after a uniform specimen temperature has been reached. Measurements at intermediate temperatures would not be accurate because of nonuniform cooling of the specimen during precooling with liquid nitrogen. After blowing out any remaining liquid nitrogen, cooling with liquid helium is started and an accurate measurement of specimen contraction made at the lowest temperature after thermal stability is reached. At this time, the recording equipment can be rezeroed and the curve of temperature vs. expansion recorded as the specimen is heated from the lowest temperature to room temperature using the heater. The two readings obtained during cooling can be used to check the heating curve.

With some materials it may be necessary to start recording the specimen length change from room temperature down and back to room temperature at a uniform cooling and heating rate. After purging and back filling with helium gas, the specimen should be cooled with liquid helium only, without liquid nitrogen precooling. This procedure avoids freezing and eliminates the disruption in cooling rate that would result from using liquid nitrogen.

Comment

The author would like to point out that Styrofoam insulation is not as good as vacuum. The primary reason for using Styrofoam in the cryostat is to offer an alternative for vacuum so that a vacuum pump is eliminated. Also, Styrofoam has been used successfully in a tensile testing cryostat.

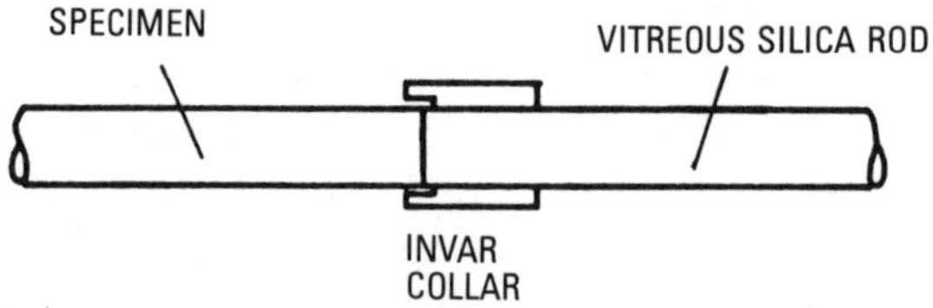

Note: Cement Invar Collar to Vitreous Silica Rod Using Micro-Measurements Inc. M-Bond 610

FIGURE 11–2B.—Invar collar cemented to vitreous silica rod

N. B. S. Cryogenic Dilatometer

The dilatometer described in this section is used at the National Bureau of Standards, Boulder, Colorado. It has been described in references 1 and 2. The dilatometer is shown in Figure 11-3.

An 8-inch-long specimen is placed at the bottom of a long vitreous silica tube. The change in length of the specimen is transmitted to a dial gage by a vitreous silica tube rider. The dial gage is covered with a bell jar and maintained in a vacuum at room temperature. A vacuum space separates the concentric tubes from the cryogenic liquid. A helium exchange gas pressure of 1.5 PSIG is maintained in the sample chamber at all times. The helium exchange gas cools the specimen after the apparatus is immersed in a cryostat containing liquid nitrogen or liquid hydrogen. The temperature of the specimen is measured with chromel-constantan thermocouples attached directly to the specimen. A cooling rate of 1-2°C (1-2°K) per minute was chosen as the best compromise between the speed of measurement and minimal temperature gradients in the specimen in order to allow the determination of the average temperature to ¼°C or better. The apparatus is calibrated by measuring the thermal expansion of polycrystalline OFHC copper.

European Organization for Nuclear Research Cryogenic Dilatometer

The cryogenic vitreous silica dilatometer built and used by the European Organization for Nuclear Research, Geneva, Switzerland, was previously discussed in reference 3 and is shown in Figure 11-4. This dilatometer is used primarily to test composite materials for the European nuclear industry to determine the influence of parameters such as fiber and resin types, fiber configuration, fiber orientation, and fiber-to-resin ratios.

The dilatometer has an outer tube to hold the specimen and an inner tube whose motion is measured relative to the outer tube as the specimen length changes during cooling. The inner tube is made with two spheres which prevent lateral motion and guide the tube vertically. The specimen, two inches long, is positioned at the bottom of the outer tube by means of vitreous silica balls fitting into a spherical hollow in the center of each end of the specimen. The cryogenic liquid flows around the specimen through a window cut

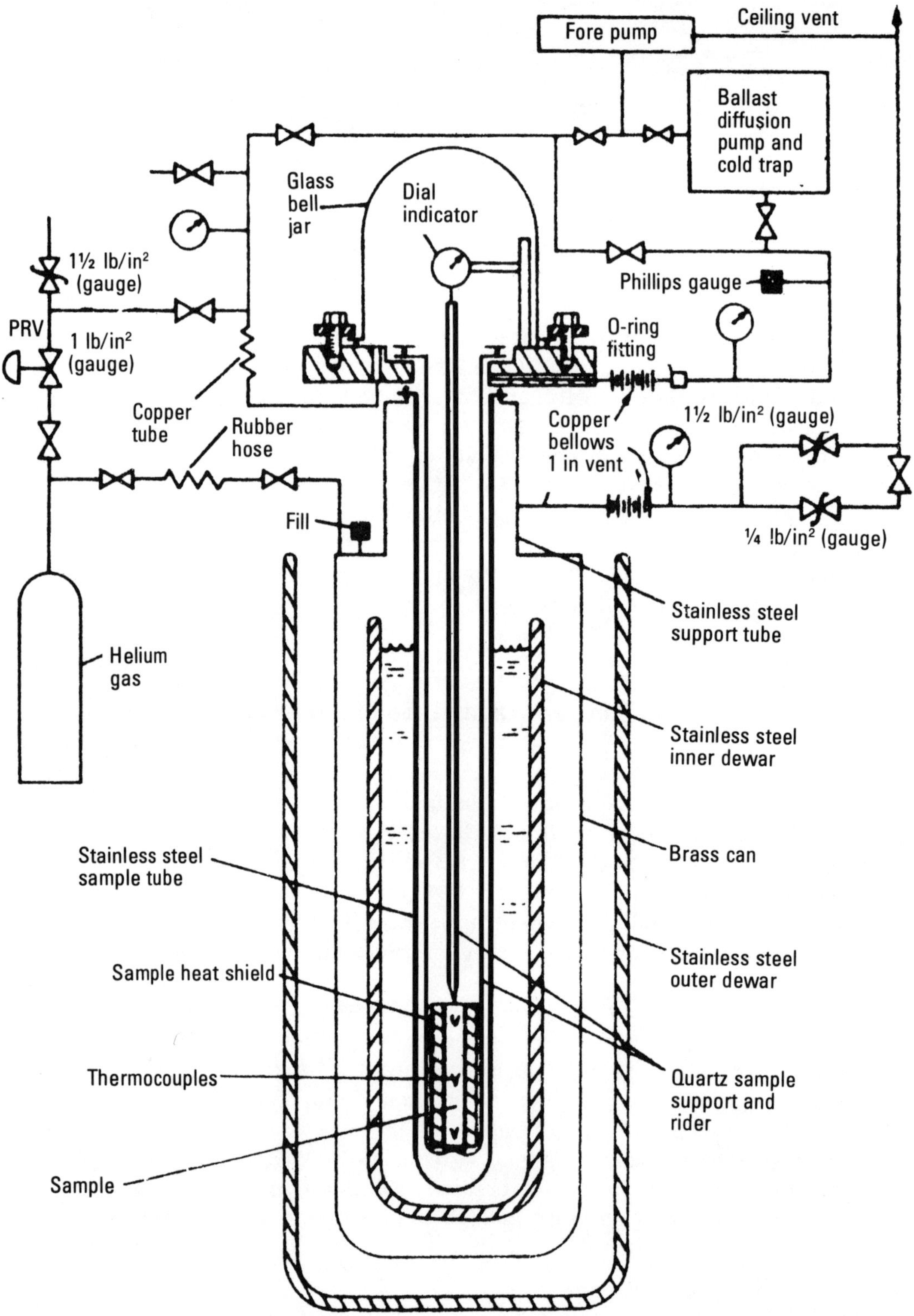

FIGURE 11–3.—Diagram of N.B.S. dilatometer.

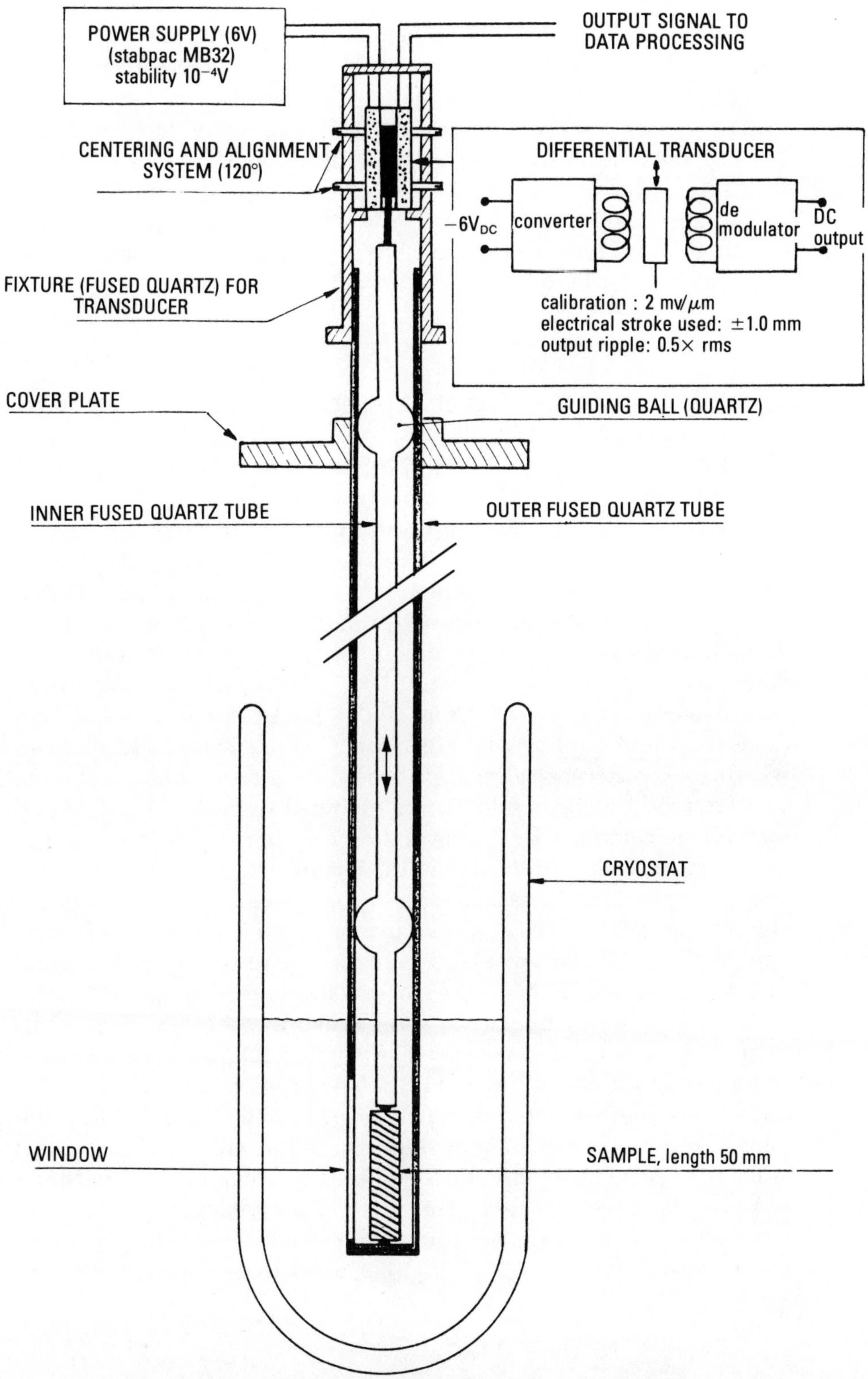

FIGURE 11–4.—European Organization for Nuclear Research cryogenic dilatometer.

into the outer tube at the specimen end so that the specimen is completely immersed in the cryogenic fluid. The change in specimen length is measured with an LVDT fixed at the open end of the outer tube.

The system is calibrated at −196°C (77°K) using liquid nitrogen and at −269°C (4.2°K) using liquid helium to cool the system. The tests are started at room temperature. The absolute accuracy of the measurement is reported to be 79×10^{-6} in/in (2 μm). Since the system is very sensitive to vibrations, a special shock-absorbing fixture was developed to hold the cryostat.

French Cryogenic Differential Dilatometer

A low temperature −269°C to 77°C (4.2°K to 350°K) vertical vitreous silica tube differential dilatometer was described by C.A.V. de A. Rodrigues during the Seventh International Thermal Expansion Symposium held November, 1979, at I. I. T. Research Institute of Chicago, Illinois. This dilatometer is illustrated in Figure 11-5A. Thermal dilation is measured with an inductive displacement transducer having a linear range of ±1 mm. At present, the apparatus can detect a change of 1×10^{-4} mm (4×10^{-6} inches) for a maximum length change of $\pm\ 3 \times 10^{-1}$ mm (0.012 inches) or 3.33×10^{-4} mm (13×10^{-6} inches) for a length change of ±1 mm (.040 inches), which makes it especially useful to study phase transformations. The force on the specimen can be changed from 0 to 1N during a test. Linear cooling and heating of the specimen and the reference at rates up to 5°K per minute are attained using a liquid helium cryostat and a furnace with its associated electronic equipment. The sample can also be held at any temperature between 4.2°K to 350°K. The apparatus has been used to study austenite to marsenite transformation phenomena of FeNiC alloys which have a low Ms temperature.

The specimen temperature is regulated by a heater which is attached to the bottom of the metal protection tube in the area of the specimen as shown in Figure 11-5B. The protection tube covers the entire length of the vitreous silica tube. In addition to holding the heater, the protection tube serves as a baffle around the vitreous silica tube preventing direct contact of the cryogenic fluid with the sample and the reference. During a test, the cryogenic fluid flows to the bottom of the cryostat from a transfer tube having an inside diameter of about 1 mm (.039 inch). A pressure is imposed upon the

A. GENERAL VIEW

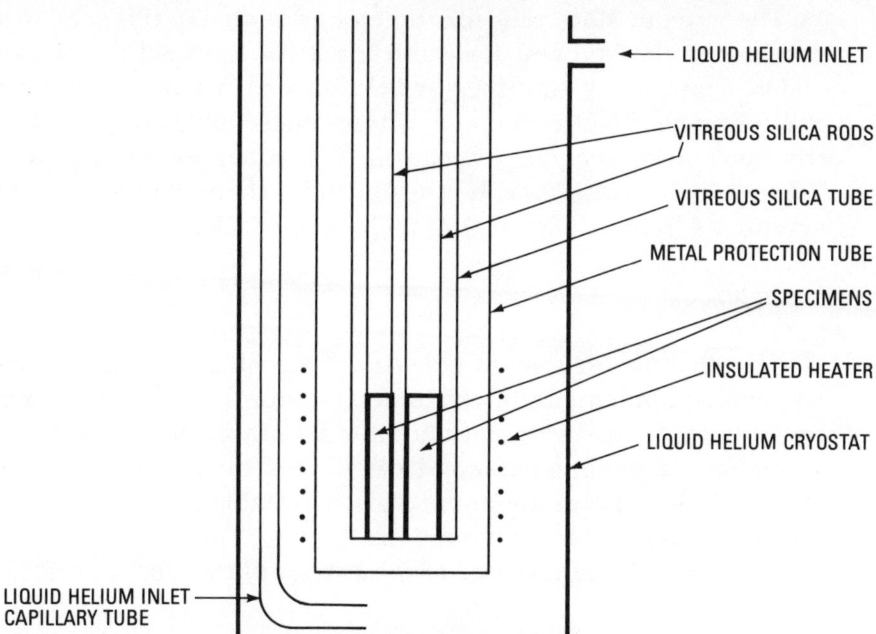

B. SPECIMEN AREA DETAIL

FIGURE 11–5.—French cryogenic differential dilatometer.

liquid helium in the storage dewar to attain a proper flow rate into the cryostat. As the liquid helium vaporizes at the bottom of the cryostat and cools the specimen, the power to the heater is controlled with a program controller to attain a specified cooling or heating rate. By balancing the power to the heater with the helium flow rate, the specimen temperature can be held constant. The measuring portion of this dilatometer is not shown because the author does not have these details. The measuring section of the other differential dilatometers described in this book can be used with the apparatus discussed in this section. Also it is the intent of the author to show the low temperature components and the techniques of temperature control in the cryogenic range rather than the measuring elements, since they are generally the same as those used above room temperature.

The reader can contact the designer at the following address to obtain more information about this dilatometer: C.A.V. de A. Rodrigues, Institute de Physique et de Metallurgia, Ecole Centrale des Arts of Manufactures, 92290 Chatenay-Malabry, France. This dilatometer is for sale by the designer.

Midwest Research Institute Cryogenic Dilatometer

The vitreous silica tube dilatometer presented in this section was designed at the Midwest Research Institute, Kansas City, Missouri, and was previously described by B.L. Rhodes et al at the 1962 Cryogenic Engineering Conference, whose proceedings are published in the book *Advances in Cryogenic Engineering* distributed by Plenum Press, New York. It is designed for thermal expansion measurements from −255°C to 300°C (18°K to 573°K).

Apparatus

The dilatometer system consists of suitable cryogenic dewars, vacuum equipment, an interferometer, automatic temperature control system, temperature-measurement equipment, and a hoist for assembly of the dilatometer and specimen. The dewars, interferometer, and dilatometer are supported on a stable platform to reduce the effects of mechanical vibrations.

The general arrangement of the fused-quartz tube, interferometer head, and specimen is shown in Figure 11-6. The 10.16-cm-long by 6.2-mm-diameter specimen rests on a fused-quartz centering bead in the base of the tube. A heater and sheath surround the specimen to heat it uniformly and provide centering relative to the fused-quartz bead and sensing rod. A side opening near the base of the tube permits convenient access to the specimen and provides entrance for the heater and thermocouple leads. A long fused-quartz rod rests on top of the specimen and transmits the changes in specimen length to the interferometer. A metal guide tube with internal knife edges is provided to center the sensing rod and minimize the effects of rod vibration and buckling. The interferometer head design incorporates a precision piston and cylinder which permit measurements of linear excursions to more than 5mm. With the close tolerance of the mating surfaces, fringe-spacing changes due to piston wobble are negligible. The lower Fizeau plate rests on the precision piston, and the upper Fizeau plate rests on a spring-opposed leveling disc supported by the piston mounting frame. As shown in Figure 11-7 the interferometer telescope contains a mercury-arc lamp, a 5461 Å filter, and suitable lenses and apertures.

When the Fiseau plates are properly adjusted, a fringe pattern is focused on the eyepiece and the photomultiplier tube. The passage of fringes across the photomultiplier tube aperture produces current variations which are amplified by a photometer circuit and recorded on a single-speed strip-chart recorder. Total fringes are automatically counted by signals from a small microswitch tripped by the recorder pen; partial fringe counts are estimated from observation of the chart.

The specimen temperature is automatically controlled with a circuit containing an L & N C.A.T. controller, 9834 dc null detector, and magnetic amplifier, with a copper vs. constantan thermocouple, attached mechanically to the specimen heater. In principle, the automatic temperature control circuit maintains the specimen surface temperature at a constant value.

The specimen temperature is determined with an L & N, K-3 potentiometer and dc null detector and a copper-constantan thermocouple imbedded in the side of the specimen at its midpoint.

Operating Procedure

The system performance is checked at room temperature to ascertain proper piston motion, good temperature control and measurement, and correct fringe counting. Liquid nitrogen is placed in

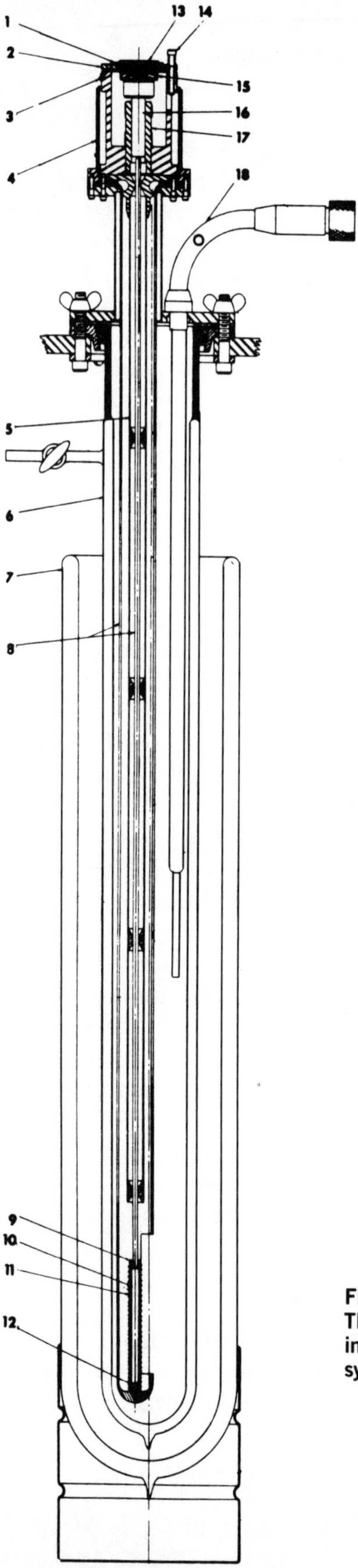

1 RETAINER RING
2 LEVELING DISC
3 LEVELING SCREWS
4 ADJUSTABLE LIGHT SHIELD
5 SENSING-ROD GUIDE TUBE
6 LIQUID-HELIUM DEWAR
7 LIQUID-NITROGEN DEWAR
8 FUSED-QUARTZ DILATOMETER AND SENSING ROD
9 GUIDE RING
10 THERMAL-EXPANSION SPECIMEN
11 SPECIMEN HEATER
12 FUSED-QUARTZ CENTERING BEAD
13 UPPER FIZEAU PLATE
14 TENSION SCREWS
15 LOWER FIZEAU PLATE
16 PRECISION PISTON
17 PRECISION CYLINDER; AND
18 TRANSFER TUBE

FIGURE 11–6.—Midwest Research Institute Thermal expansion interferometer-dilatometer system and dewar assembly.

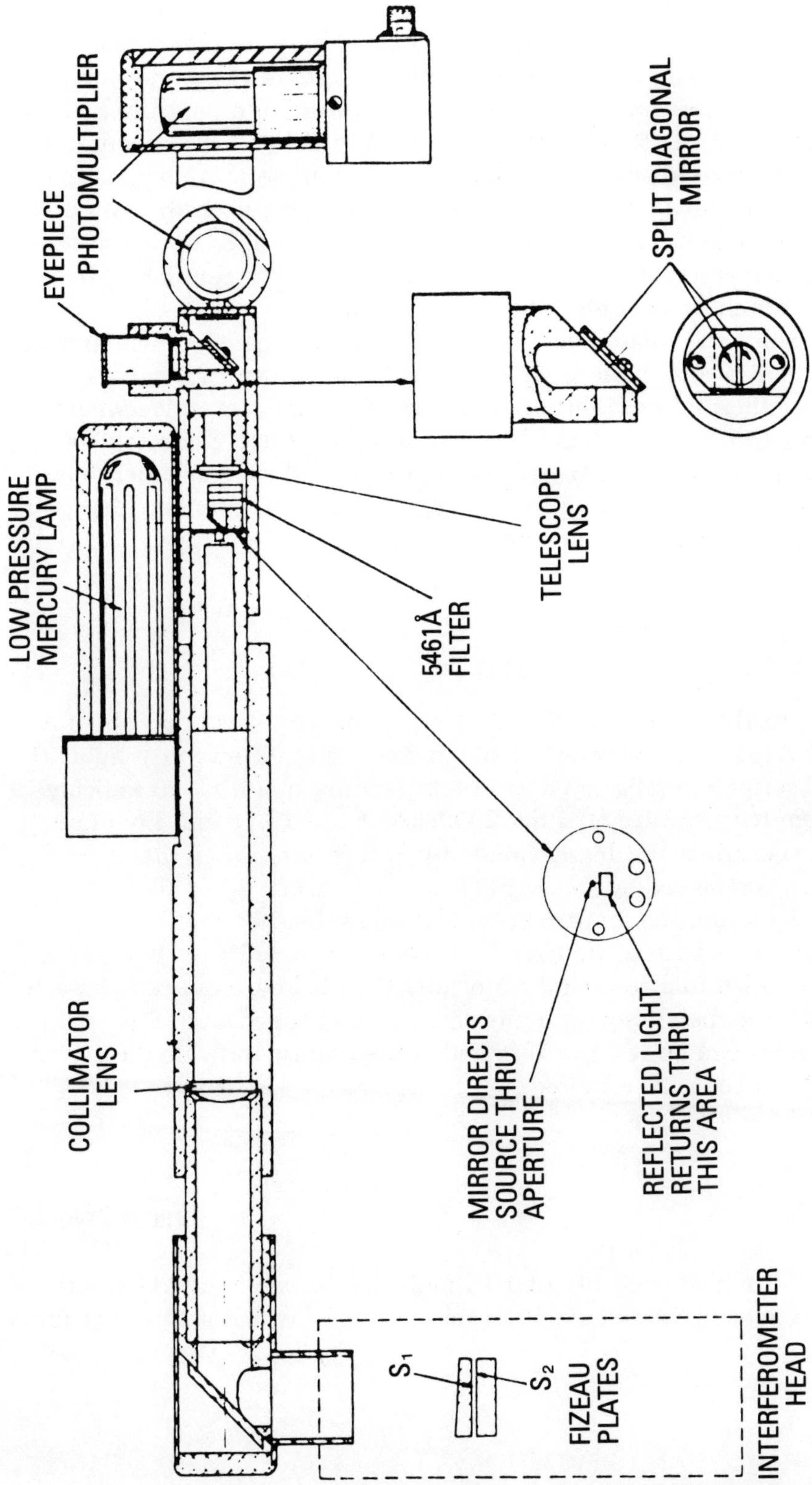

FIGURE 11–7.—Interferometer telescope and fringe detecting components.

the outer dewar, and an exchange gas is used in the inner dewar to cool the specimen to − 190°C (83°K). For measurements below this temperature liquid helium is transferred into the inner-dewar cavity until the specimen has reached a uniform temperature slightly below −253°C (20°K). All controls and initial data are recorded and the sample expansion is observed as the temperature begins to rise as the result of natural heat leaks or heating with the automatic temperature control circuit. After a new stable temperature is reached and maintained for several minutes, the number of fringes and the thermocouple potential are recorded.

Since the dilatometer and specimen are heated slowly and the temperature is held at each selected temperature for several minutes, thermal equilibrium is assumed to exist at adjacent points throughout the system. The measurements are corrected for the fused quartz tube adjacent to, and equal to, the specimen length. Under these conditions the relation of the dilatometer change to fringes recorded is given by

$$\left(\frac{L_T - L_{20}}{L_{20}}\right)_S - \left(\frac{L_T - L_{20}}{L_{20}}\right)_Q = \frac{\lambda/2}{L_{20}}(N_T - N_{20})$$

$$= 0.2688 \cdot 10^{-5}(N_T - N_{20}) \qquad (1)$$

where the subscripts S and Q represent the specimen and fused quartz, respectively, λ = 5461 Å for a filtered mercury light, the subscript T represents a local temperature quantity, 20 indicates a room temperature quantity 20°C, and L_{20} = 10.16 cm. Equation (1) also describes the decreasing-temperature data, when the $(N_t - N_{20})$ term is replaced by $(N_{20} - N_t)$.

The fringe counts are accumulated as the specimen temperature is increased from the initial temperature to 300°C or is decreased from an initial temperature of 300°C to below room temperature. The specimen temperature is normally controlled at 20°C so that the counter can be reset to zero and subsequently indicate directly the difference count required in (1).

Janis Cryostat

A sophisticated cryostat suitable for a cryogenic dilatometer is described in this section. It is commercially available from the Janis

Research Company, Inc., of Stoneham, Massachusetts. With the use of this kind of cryostat, identified by the manufacturer as the Super Varitemp research cryostat, the most important element of the cryogenic dilatometer is complete. All that one needs is the measuring element, which is also available, since the same measuring elements used in elevated temperature dilatometers can be employed in the cryogenic dilatometer.

The features of the cryostat and its description are presented. The basic operation of a Super Varitemp cryogenic system is also discussed to give the reader a more detailed procedure to appreciate better the techniques of testing in the cryogenic temperature range, which is a new area of interest for most researchers.

Features of the Cryostat

Details of a typical variable temperature cryostat are shown in Figure 11-8, and a photograph of a typical cryostat is shown in Figure 11-9.

The temperature range of the cryostat is −271°C (1.5°K) to 25°C (298°K). A specimen can be changed without temperature-or vacuum-cycling the cryostat, and a specimen can be cooled from room temperature to below 269°C (4°K) in less than five minutes. The cryostats can be operated efficiently at low temperatures continuously. Only the specimen chamber is pumped, leaving the main helium reservoir at 269°C (4.2°K). The tail of the cryostat is detachable and can be replaced with others for additional applications. Optical access both in the vertical and/or horizontal axis is available. Automatic temperature control is available to heat and cool the specimen at a constant rate over the range 0°K to 298°K.

Description

The central tube of the dewar functions as the sample chamber, with access being provided from the top. Suitable pump-out or venting ports are provided for the cooling vapor at the top of this assembly. A unique throttling valve, located in the helium reservoir, is connected to the bottom of the sample chamber through a length of capillary tubing. Liquid is drawn from the main reservoir, through the valve and capillary, and is introduced into the bottom of the sample chamber through a diffuser. The primary function of the diffuser is to break up the entering gas stream thus providing a

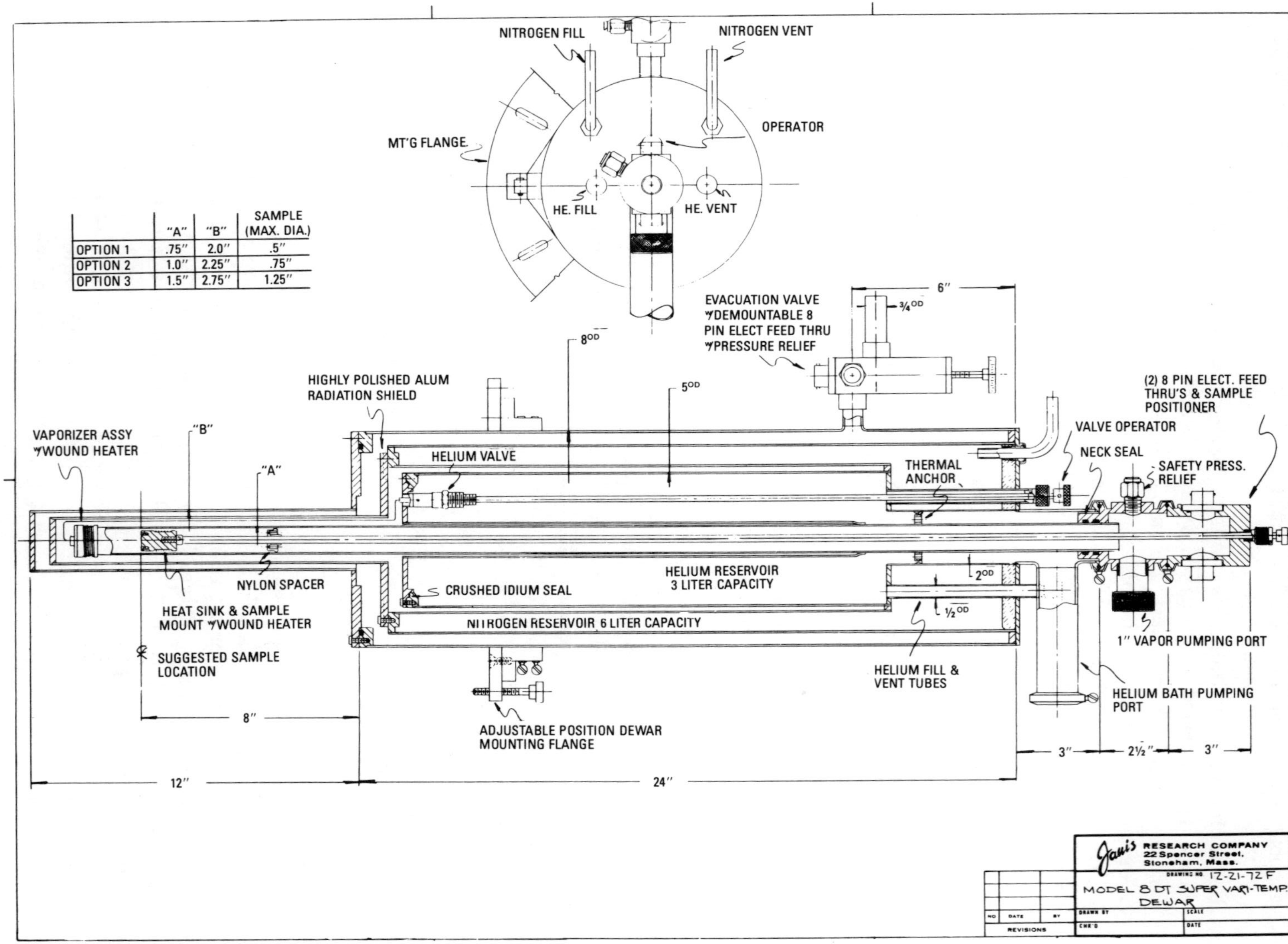

	"A"	"B"	SAMPLE (MAX. DIA.)
OPTION 1	.75"	2.0"	.5"
OPTION 2	1.0"	2.25"	.75"
OPTION 3	1.5"	2.75"	1.25"

FIGURE 11–8.—Typical Janis Super Varitemp cryostat detail.

FIGURE 11–9.—Photograph of typical Janis cryostat.

uniform fluid profile across the tube. Vapor flow is then upward, around the sample, and out the vent port at the top of the assembly. For temperatures above 4.2°K, venting is to the atmosphere. For temperatures below 4.2°K the sample chamber is pumped with a small mechanical pump. Elevated sample temperatures are obtained by heating either the diffuser for rought control of vapor temperature or an additional sample heater for more precise regulation. Optimum control is achieved by using a combination of both, where partial regulation is accomplished by heating the entering gas stream to approximately the desired temperature; and precise regulation is obtained with the sample heater, usually with an automatic temperature controller. In practice the helium throttling valve is set to provide the necessary fluid flow (rate of cooling), and all further temperature regulation is accomplished electronically.

Basic Operation of the Super Varitemp System

Following is a step-by-step procedure for the use of a Super Varitemp System in testing from about 1.5°K to 298°K. Refer to Figures 11-10 and 11-11 for the elements of the system.

1) Attach a mechanical vacuum pump to the vapor pumping port (A) at the top of the cryostat. Continuously evacuate the sample tube (B).

2) Precool the cryostat with liquid nitrogen. The helium reservoir may be precooled with liquid nitrogen or back filled with helium gas.

3) If liquid nitrogen is used to precool the helium reservoir, blow out using helium gas and leave the reservoir back filled (with helium gas).

4) Open the helium throttle valve control (C) approximately one turn and listen for the "gurgle" from the vacuum pump. This indicates that the valve (D), the helium capillary (E), and the vaporizer (F) are clean and open. The entire sample tube (B) will then be purged with helium gas. Upon purging the system, close the helium valve finger tight. (Do not overtighten or the valve may freeze shut as the result of the contraction of the dis-similar materials used in the valve body and needle assembly.)

5) Fill the helium reservoir to the desired volume of liquid helium. Upon initiation of helium transfer, repeat step (4). It is a good idea to continue to repeat step (4) several times during the filling operation to insure that the entire "Super Varitemp" system remains clean and open to proper helium flow.

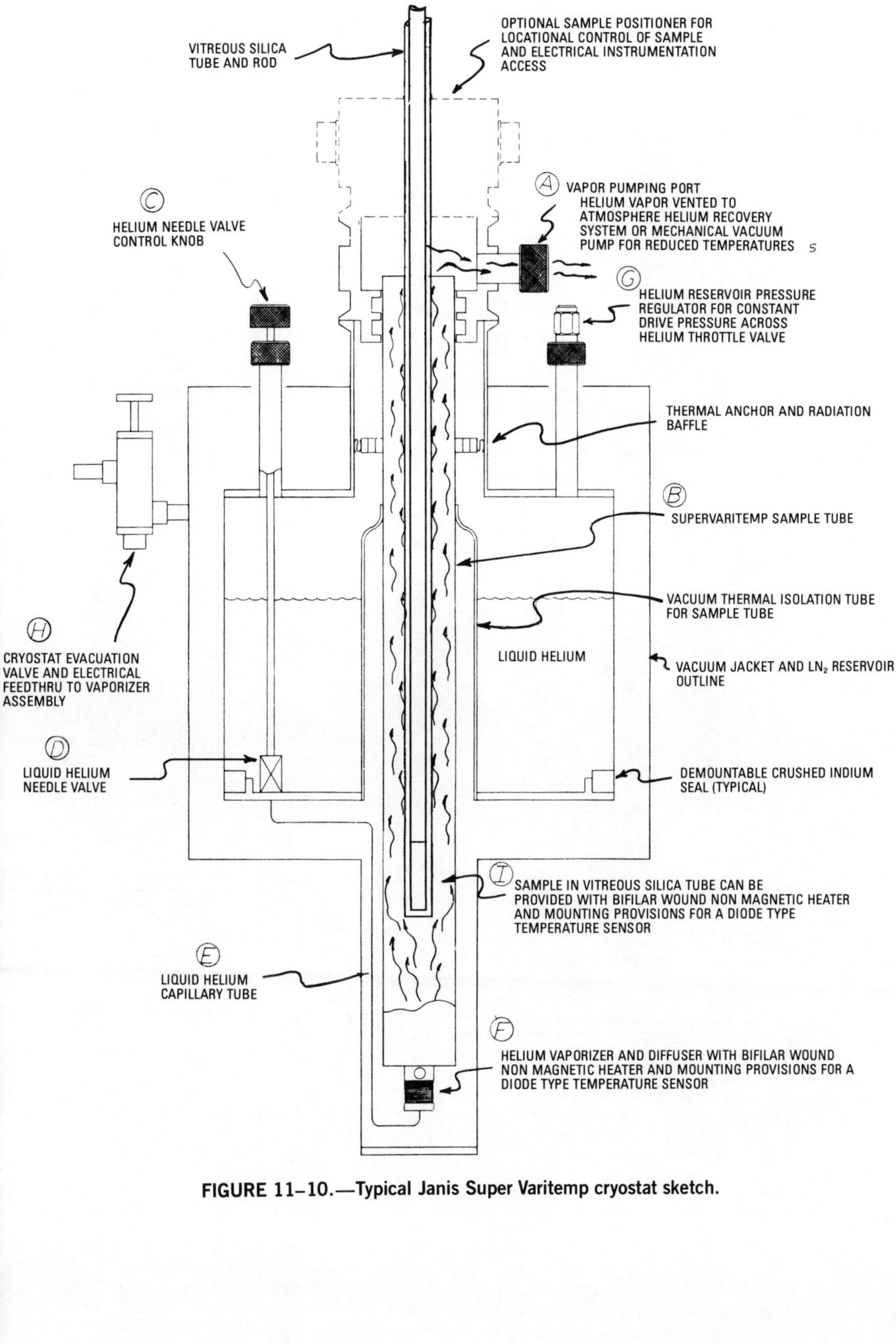

FIGURE 11–10.—Typical Janis Super Varitemp cryostat sketch.

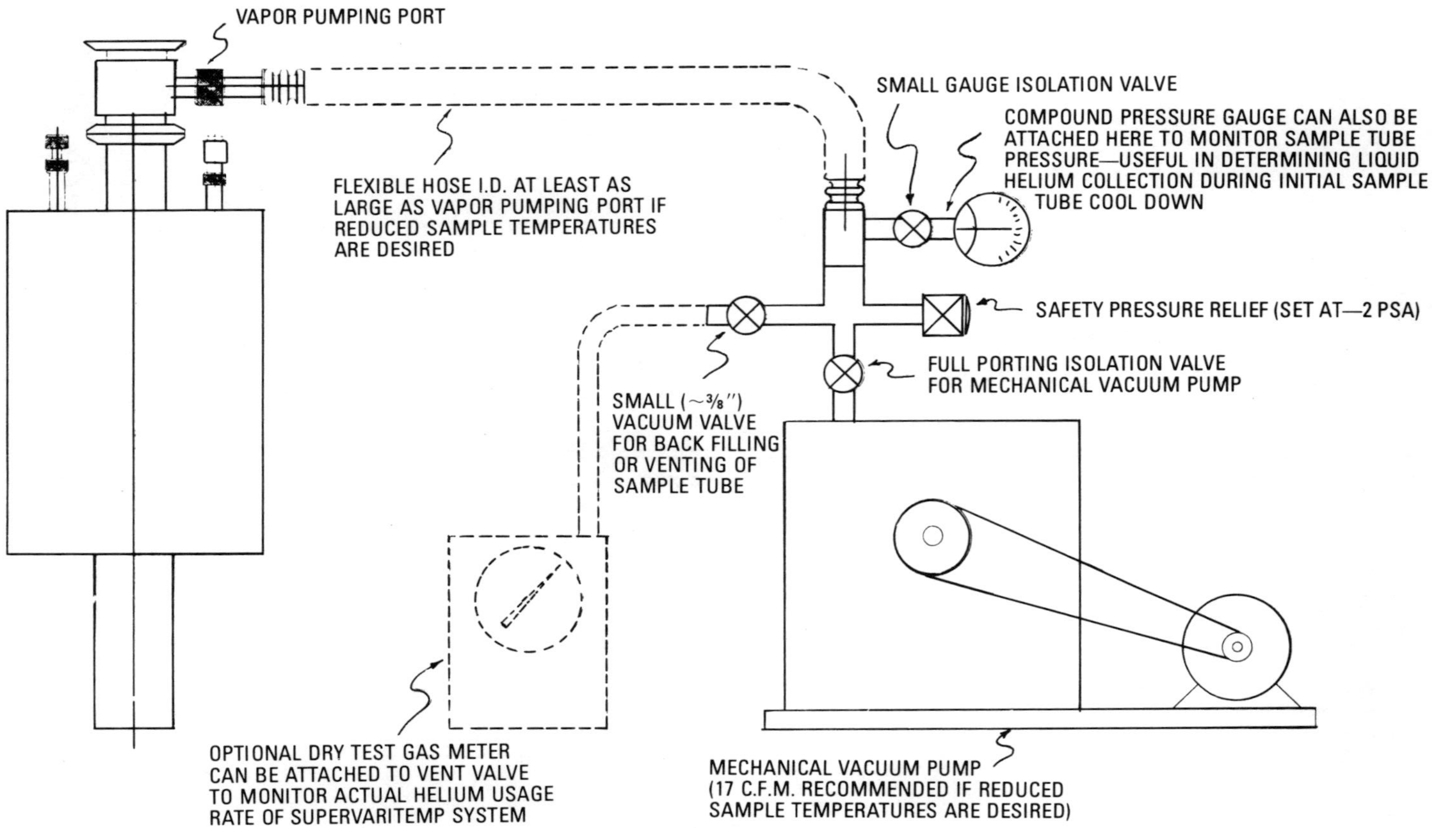

FIGURE 11–11.—Manifold Assembly for Super Varitemp cryostat.

6) Upon completion of the liquid helium transfer, install the helium reservoir pressure regulator (G). Be sure other access ports to the reservoir are closed and sealed. The regulator will then maintain a positive pressure (2 to 4 psig, depending upon the system) which both precludes the admission of air to the helium reservoir and provides a constant drive pressure for the liquid helium across the helium throttle valve (D).

7) Once the helium reservoir is pressurized, the mechanical vacuum pump should be valved off (see the suggested manifold assembly shown in Figure 11-11) and the helium valve control (C) opened anywhere from two to six turns. The vaporizer assembly (F) will cool down and liquid helium should collect at the bottom of the sample tube within a few minutes (depending upon valve opening, sample tube size, and mass of flanges and window mounts involved).

8) Once liquid helium collection has been accomplished, the helium valve control (C) should be reclosed to approximately 90° to 120° (of one turn) open, and either a vaporizer heater power supply or an automatic temperature controller should be attached to the electrical feedthrough (H) on the evacuation valve. It is important to note that once the pressure in the sample tube (B) exceeds atmospheric pressure, it must be vented to either atmosphere or to a helium recovery system. Otherwise the back pressure in the tube could preclude the actual collection of liquid helium. An excellent way of monitoring this pressure is to install a compound pressure gage in the valve manifold assembly (see Figure 11-11) at the mechanical pump.

9) The "Super Varitemp" system is now ready to operate. For normal efficiency the vaporizer heater current should be set to approximately 50-75 ma and the helium valve control (C) adjusted so that the temperature in the sample zone (I) is approximately −265°C to −263°C (8°K to 10°K). These values will vary depending upon the particular system parameters involved, but can be considered a good "rule of thumb" to start from. At this point, temperature can be varied from 5°K to 298°K, merely by changing the vaporizer heater power. It is not necessary to adjust the helium valve control (C) again unless maximum efficiency of helium usage is desired.

10a) To obtain temperatures from below −271°C (2°K) to approximately −253°C (20°K), with the sample in vapor, close the helium valve control (C), and the vent valve of the manifold assembly (Figure 11-11), and open fully the mechanical vacuum pump isolation valve. Once the sample tube (B) is evacuated, the helium valve control (C) should be slightly opened and adjusted until a pressure of approximately 3 to 5 TORR can be maintained in the sample tube (B) with a vaporizer heater current of approximately 100 to 150 ma. The temperature of the sample zone (I) can then be adjusted by vaporizer heater power alone.

10b) To obtain temperatures from below 2°K to 4°K, with sample in liquid helium, turn off the vaporizer heater, open fully the helium valve control (C), and flood the sample tube (B) with liquid helium. The helium valve can then be either closed and the liquid helium pumped to the lowest possible temperature (vent valve closed and pump isolation valve open), or throttled in order to replenish continuously the liquid helium in the sample tube. The latter method naturally limits somewhat the lowest achievable temperature.

11) In order to change samples in the "Super Varitemp" system (temperatures of 4°K and above) it is only necessary to remove the sample holder from the top of the sample tube and then reinsert the holder with the new sample. During this time the helium vapor flow in the sample tube should be maintained, or even increased, in order to preclude the admission of any air to the sample zone. When operating below 4°K one should bring the temperature to 4°K and the pressure to 760 TORR before removing the sample holder.

12) Shutting down procedures for the "Super Varitemp" system are quite straight-forward. The helium valve control (C) should be closed, the vaporizer heater power turned off, and ideally, the sample tube (B) continuously evacuated until the entire system has returned to room temperature. If this presents a problem, the sample tube can just be valved off (close all valves shown in the manifold assembly of Figure 11-11) until the entire system reaches room temperature. All "Super Varitemp" sample tubes are protected with an overpressure relief valve to avoid dangerous pressure increases due to evaporating liquid left in the sample tube.

13) Warning: The vaporizer heater (F) must not be energized when there is no helium flow present. The extreme thermal isolation achieved through the basic system design will allow the vaporizer to quickly reach destructive temperatures and either burn out the heater or melt the soft solder joint commonly used at the bottom of the assembly.

References

1. V. Arp, J. H. Wilson, L. Winrich, and P. Sikora, "Thermal Expansion of Some Engineering Materials from 20 To 293°K," *Cryogenics*, June 1962.
2. A. F. Clark, "Low Temperature Thermal Expansion of Some Metallic Alloys," Cryogenics, October, 1968.
3. K. Dahlerup-Peterson, "Tests of Composite Materials at Cryogenic Temperatures: Facilities and Results," European Organization for Nuclear Research, Geneva, Switzerland. Paper presented at 1979 International Cryogenic Materials Conference, University of Wisconsin, August 21-24, 1979.

Appendix

ANS1/ASTM E228-71 (Reapproved 1979)

Standard Test Method for Linear Thermal Expansion of Rigid Solids with a Vitreous Silica Dilatometer[1]

This Standard is issued under the fixed designation E 228; the number immediately following the designation indicates the year of original adoption or, in the case of revision, the year of last revision. A number in parentheses indicates the year of last approval.

1. Scope

1.1 This method covers the determination by a vitreous silica dilatometer of the linear thermal expansion from − 195 to + 1000 C (Section 4) for rigid solids including metals, ceramics and refrac-

1. This method is under the jurisdiction of ASTM Committee E-37 on Thermal Measurements.
Current edition effective Nov. 19, 1971. Originally issued 1963. Replaces E 228-66a T.

tories, glasses, rocks and minerals, plastics, wood, and inorganic cements, pastes, and mortars.

1.2 For this purpose, a rigid solid is defined as a material which, at test temperatures and under the stresses imposed by specimen-supporting members in the thermal expansion apparatus, has a negligible creep or elastic strain rate, or both, insofar as they would significantly affect the precision of thermal length change measurements.

1.3 It is recognized that many rigid solids require detailed preconditioning and specific thermal test schedules for correct evaluation of linear thermal expansion behavior for certain materials applications. Since a general method of test cannot cover all specific requirements, details of this nature should be discussed in particular material specifications.

Note 1—The values stated in U.S. customary units are to be regarded as the standard.

2. Description of Terms

2.1 *Linear Thermal Expansion* is the change in length per unit length resulting from a change in temperature of the material. Symbolically represented by $\Delta L/L_0$, where ΔL is the observed change in length and L_0 is the length of the specimen at reference temperature T_0, linear thermal expansion has the units of inches per inch, or centimeters per centimeter, often expressed as percentage or parts per million.

2.2 *Mean Coefficient of Linear Thermal Expansion*, α_m, between temperatures T_1 and T_2, is defined as:

$$\alpha_m = (L_2 - L_1)/L_0(T_2 - T_1) = \Delta L/(L_0 \cdot \Delta T)$$

where L_1 and L_2 = specimen lengths at temperatures T_1 and T_2, respectively. α_m is therefore obtained by dividing the linear thermal expansion ($\Delta L/L_0$) by the change of temperature (ΔT). Units are inches per inch, or centimeter per centimeter per degree change in temperature, often expressed in parts per million per degree.

2.3 *Instantaneous Coefficient of Linear Thermal Expansion*, α_T, at temperature T, is defined by the following expression:

$$\alpha_T = \underset{T_1 \to T_2}{\text{Limit}} (L_2 - L_1)/L_0(T_2 - T_1) = dL/(dT \cdot L_0)$$

α_T has the same units as α_m.

3. Apparatus

3.1 *Dilatometer*, of the tube or rod type, fabricated of high-purity vitreous silica. Typical dilatometers of these two types are shown in Appendix Figs. 1 and 2.

3.2 *Extensometer*, for measuring length changes over a length of at least 0.050 in. (1.27 mm). The graduated scale shall permit direct reading to 0.0001 in. (0.003 mm) and estimates of 0.00002 in. (0.0005 mm). Its accuracy shall be such that the error of indication will not exceed ± 0.00005 in. (0.0013 mm) for any length change. To ensure this accuracy, the extensometer shall be precalibrated and its mounting assembly shall be protected from excessive temperature changes. Invar parts are recommended to eliminate ambient temperature effects (see Appendix Figs. 1 and 2). Dial gages are commonly used, but any device (magnetic, capacitative, optical, or other) may be used provided that length changes are reliably measured as specified. Extensometers that exhibit instability with time are not recommended and, if used, shall be frequently calibrated (see A1, page 202).

3.3 *Micrometer Calipers*, with an index permitting direct reading of 0.001 in. (0.025 mm) for measuring the initial specimen length. A high-grade screw micrometer customarily used in machine shop practice is satisfactory.

3.4 *Electric Furnace*, capable of maintaining the difference between the maximum and minimum temperatures of the specimen within 2 C. The details of typical furnaces are shown in Appendix Figs. 3 and 4. Temperature-control equipment shall maintain the mean temperature of the specimen within ±2 C of the desired test temperature.

3.5 *Liquid Baths* may be used when expansion data below 100 C are required. The bath shall be arranged so that a uniform temperature throughout the specimen is maintained. Means to control the desired temperature to within ± 0.2 C shall be provided.

Note 2—If it is undesirable to have the specimen in contact with the liquid, the tube-type dilatometer shall be used without a hole for the thermocouple, and a thermometer used for temperature measurement.

3.6 *Temperature-Measuring Instruments*[2]—A calibrated thermocouple shall be provided for determining the temperature of

2. For further information, see ASTM Method E 220, for Calibration of Thermocouples by Comparison Techniques, which appears in the *Annual Book of ASTM Standards*, Parts 17 and 44.

the test specimen. Type T, E, S, K, or J thermocouples may be used. Types T and E can be calibrated to indicate temperatures accurate to ±0.2 C and ±0.5 C, respectively, in the range −190 to 350 C. A Type S thermocouple can be calibrated to indicate temperatures accurate to ±0.5 C in the range 0 to 1000 C and is especially recommended for use in the range 350 to 1000 C. A Type S thermocouple should not be used for subzero temperatures. A Type K thermocouple can be calibrated to indicate temperatures accurate to ±0.5 C in the range −190 to 350 C. A Type J thermocouple can be calibrated to indicate temperatures accurate to ±0.5 C in the range 0 to 350 C. The thermocouple shall be referenced to 0 C by means of an icewater bath or its equivalent ice-point reference system and its emf measured with a calibrated potentiometer having a sensitivity of ±1 μV and an accuracy of ±5 μV. Precautions shall be taken to ensure that the reference junction is maintained at 0 C throughout the test.

4. Safety Precautions

4.1 The use of a vitreous silica dilatometer above 800 C may be accompanied by viscous flow and a time-dependent change of thermal expansion in the vitreous silica. The magnitude of these effects above 800 C will depend on the particular type of vitreous silica used to fabricate the dilatometer. To minimize errors caused by these effects, frequent calibration of the dilatometer is recommended when expansion measurements above 800 C are made.

4.2 If a silver cup is used (optional), it shall be removed from the furnace for measurements above 850 C to prevent it from melting.

5. Test Specimens

5.1 In the fabrication of test specimens several design considerations shall be followed:

5.2 Specimen length should be between 2 and 5 in. (51 and 127 mm). Generally, specimens shorter than 2 in. result in a loss of sensitivity while specimens longer than 5 in. are subject to axial temperature differences in excess of the specified 2 C because of furnace gradients.

5.3 The minimum diameter or thickness of the specimen shall be 3/16 in. (4.8 mm) or one sixteenth of the specimen length, whichever is smaller. Smaller sections may be subject to stresses large enough to produce significant creep or elastic strain rates, or both. The maximum diameter or thickness is determined by the inside diameter of the tube-type dilatometer and the distance between fixed and transmission rods in the rod-type dilatometer.

5.4 The shapes of specimen ends and the vitreous silica contact surface shall be designed so that the specimen remains laterally fixed during the test. Appendix Figure 5 illustrates some commonly accepted designs. In the case of the rod-type dilatometer, Appendix Figs. 5 (*a*) and (*b*) are recommended; all designs are suitable for the tube type. For materials that soften, Appendix Figs. 5 (*b*) and (*c*) are recommended to minimize deformation.

5.5 Conditioning of specimens is generally necessary before reproducible expansion data can be obtained. For example, heat treatments are frequently necessary to eliminate certain effects (strain, moisture, etc.) which may introduce length changes not associated with thermal expansion.

6. Verification of Furnace and Dilatometer

6.1 With the specimen-holding vitreous silica assembly in normal position in the furnace, place three calibrated thermocouples, housed in double-bore ceramic tubes and bundled together, into the furnace so that their junctions lie axially at positions closely representing the center and both ends of the specimen. Measure the temperatures at these positions during heating, cooling, and at constant-temperature conditions. The maximum allowable temperature difference between any two measuring points is 2 C as specified in 3.4. If these temperature conditions are not met, the furnace shall be modified. Generally, the use of a silver cup as shown in Appendix Fig. 3 will reduce the furnace gradient over the specimen length to within these limits.

6.2 Place in the dilatometer a specimen composed of the same vitreous silica as that used for the fabrication of the dilatometer as described in 8.1. Vary the temperature of the dilatometer over the intended operating range as described in 8.2. Record extensometer readings at 50 C intervals. The extensometer ideally should indicate no change as the temperature is varied. The maximum permissible difference between the highest and lowest extensometer readings shall be 0.00005 in. (0.0013 mm). Excessive extensometer variation during this test is indicative of mechanical or thermal instability in the apparatus and must be corrected (see A4, page 203).

7. Calibration

7.1 Determine the correction for the vitreous silica thermal expansion by heating or cooling chemically pure platinum in the dilatometer over the desired temperature range. Calculate the difference between the true linear thermal expansion of platinum and the observed differential linear thermal expansion between platinum and vitreous silica. This represents the linear thermal expansion of vitreous silica. Oxygen-free high-conductivity (OFHC) copper may be used for calibration between − 195 and + 100 C. Data for the linear thermal expansion of platinum and OFHC copper are given in A2, page 202.

8. Procedure

8.1 Clean the specimen and install it in the dilatometer after making certain that the end surfaces, as well as the contact surfaces of the dilatometer, are free of foreign particles. Take care to assure good seating of the specimen in the dilatometer. Place the thermocouple junction at the midpoint of the specimen. Either embed the junction in the specimen or place it on or within 0.5 mm of its surface (see A3, page 202). Mount the extensometer to provide good stable contact with the transmission rod and set it to a convenient reading. Insert the dilatometer assembly into the furnace or bath and then heat or cool by the furnace or baths, or both, to give thermal schedules described in 8.2 and 8.3.

8.1.1 *Procedure A*, a series of constant temperatures.

8.1.2 *Procedure B*, heating or cooling at constant rate.

Note 3—In general, Procedure A results in greater accuracy. It is more practical to use Procedure B because (*1*) a series of constant temperature holds is more time consuming and (2) in temperature regions where the expansion of the material is time dependent, constant rate conditions usually lead to easier comparison of data.

8.2 *Procedure A*—Heat or cool the dilatometer assembly between any two temperatures, leaving it at each temperature until the extensometer shows no significant change. The minimum holding time for a liquid bath shall be 20 min and for the furnace, 25 min. Read the extensometer and temperature and record. Before any reading of the extensometer is made, the dilatometer mounting may be gently tapped to eliminate frictional drag. An alternative scheme to eliminate frictional drag is to actuate a low-energy electrical buzzer mechanically fixed to the extensometer mount prior to reading. Calculate the mean coefficient of linear thermal expansion as prescribed in 9.2.

Note 4—The reliability of the measured value of mean coefficient of linear thermal expansion may be determined as given hereafter for materials with reversible thermal expansion.

8.2.1 Repeat the procedure. Calculate the acceptable difference between two consecutive determinations over any temperature interval $(T_2 - T_1)$ as:

$$\Delta\alpha_m = [A\bar{\alpha}_m + (0.00005/L_0)]/(T_2 - T_1)$$

where:

$\Delta\alpha_m$ = acceptable differences in coefficient values, in./in. · deg C,
A = permissible uncertainty in the temperature difference, $T_2 - T_1$,
A = 2 C if the furnace is used for both temperatures,
A = 1.5 C if a liquid bath is used for one temperature,
A = 1.0 C if T_2 and T_1 are both determined by liquid baths,
$\bar{\alpha}_m$ = average value for the coefficient as calculated from the two determinations, in./in. · deg C,
L_0 = reference temperature length of the specimen, in., and
T_1 and T_2 = the two temperatures between which the coefficient is determined, in degrees Celsius.

8.2.2 If a reliable check is not obtained, for example, if the repeat determination does not agree within the accepted precision, make another determination. When a reliable check is obtained, average the results of the last two determinations and report this value as the mean coefficient of linear thermal expansion for the temperature interval over which it was measured.

8.3 *Procedure B*—Starting at room temperature or some other temperature, either heat or cool the specimen at some arbitrary rate. Heating or cooling rates in excess of 200 C/h are unacceptable since large difference between measured temperature and true sample temperature may result. During heating or cooling, observe the extensometer reading and temperature. Make corrections for thermocouple lead or lag as prescribed in A3, page 202. Calculate thermal expansion as prescribed in 9.1 and report as a function of temperature.

8.4 *Refrigerated Furnace Procedure*—A furnace (see Appendix Fig. 3) equipped with a silver cup may be used for expansion measurements to −195 C. The silver cup serves as a container into which liquid nitrogen is poured to accomplish cooling. Introduce the liquid nitrogen until cooling is sufficient to maintain a full level of the liquid in the cup. This requires 3 to 4 liters of liquid nitrogen. Slowly introduce the dilatometer assembly into the furnace. When the residual liquid nitrogen boils away, the furnace will start natural heating. Control electrical power supplied to the furnace windings to maintain constant heating rate. Determine the linear thermal expansion as prescribed in 8.3. Alternative equipment that has been employed between −195 and +100 C is shown in Appendix Fig. 4.

9. Calculations

9.1 Calculate the linear thermal expansion. $\Delta L/L_0$, for each reading as follows:

$$\Delta L/L_0 = [(\Delta L)_a/L_0] + [(\Delta L)_s/L_0]$$

where:

L_0 = length of the test specimen at the reference temperature T_0,
$(\Delta L)_a$ = apparent changes in specimen length from the test temperature T_1 to the test temperature T_2, and
$(\Delta L)_s/L_0$ = change in length per unit length of vitreous silica from T_1 to T_2 as determined in Section 7.

9.2 Calculate the mean coefficient of linear thermal expansion, α_m, as follows:

$$\alpha_m = \Delta L/L_0\Delta T = [(\Delta L)_a/L_0\Delta T] + [(\Delta L)_s/L_0\Delta T]$$

where ΔT = temperature difference between the test temperatures T_1 and T_2.

9.3 Calculate the instantaneous coefficient of linear thermal expansion, α_T, at any temperature T, as follows:

$$\alpha_T = (dL/dT)/L_0$$

where dL/dT = slope of the length-versus-temperature curve at temperature T. The slope may be determined by either graphical or mathematical analysis.

10. Report

10.1 The report shall include the following:

10.1.1 Description of material, including name of manufacturer, chemical composition, thermal and mechanical history,

10.1.2 Method of preparation of test specimen, axis orientation if material is anisotropic, and thermal, mechanical, moisture, or other conditioning,

10.1.3 Form and dimensions of test specimen, including initial length L_0, and reference temperature T_0,

10.1.4 Brief description of the apparatus, including displacement and temperature measuring systems, estimate of precision, heating and cooling rates, and temperature controls,

10.1.5 Tabulation of data showing linear thermal expansion, test temperatures, and values for mean coefficients of linear thermal expansion for selected temperature intervals.

10.1.6 The following curves as required:

$$\Delta L/L_0 \text{ versus } T; \alpha_m \text{ versus } T; \alpha_T \text{ versus } T$$

where α_m should be computed from a common reference temperature to T,

10.1.7 Complete description of any unusual behavior of the specimen, such as permanent change in specimen length at reference temperature following test, excessive oxidation, scaling, discoloration, deformation, cracking, spalling, etc., which may be of value in interpreting test results, and

10.1.8 Any additional information required by a particular material specification.

References

High-Temperature Range:

1. J. B. Austin, "A Vacuum Apparatus for Measuring Thermal Expansion at Elevated Temperatures with Measurements on Platinum, Gold, Magnesium, and Zinc," *Physics*, PHYCA, Vol. 3, 1932, pp. 240-267. (It is important that Austin's preheat treatment be followed before the expansion test.)

Low-Temperature Range:

2. F. C. Nix, and D. McNair, "The Thermal Expansion of Pure Metals," *Physical Review*, PHRVA, Vol. 61, 1942, pp. 74-78.
3. F. C. Nix, and D. McNair, "The Thermal Expansion of Pure Metals," *Physical Review*, PHRVA, Vol. 60, 1941, p. 597.
4. T. H. Rubin, W. Altman, and H. L. Johnston, "Coefficients of Thermal Expansion of Solids at Low Temperatures," *Journal of the American Chemical Society*, JACSA, Vol. 76, 1954, p. 5289.

ASTM E 228

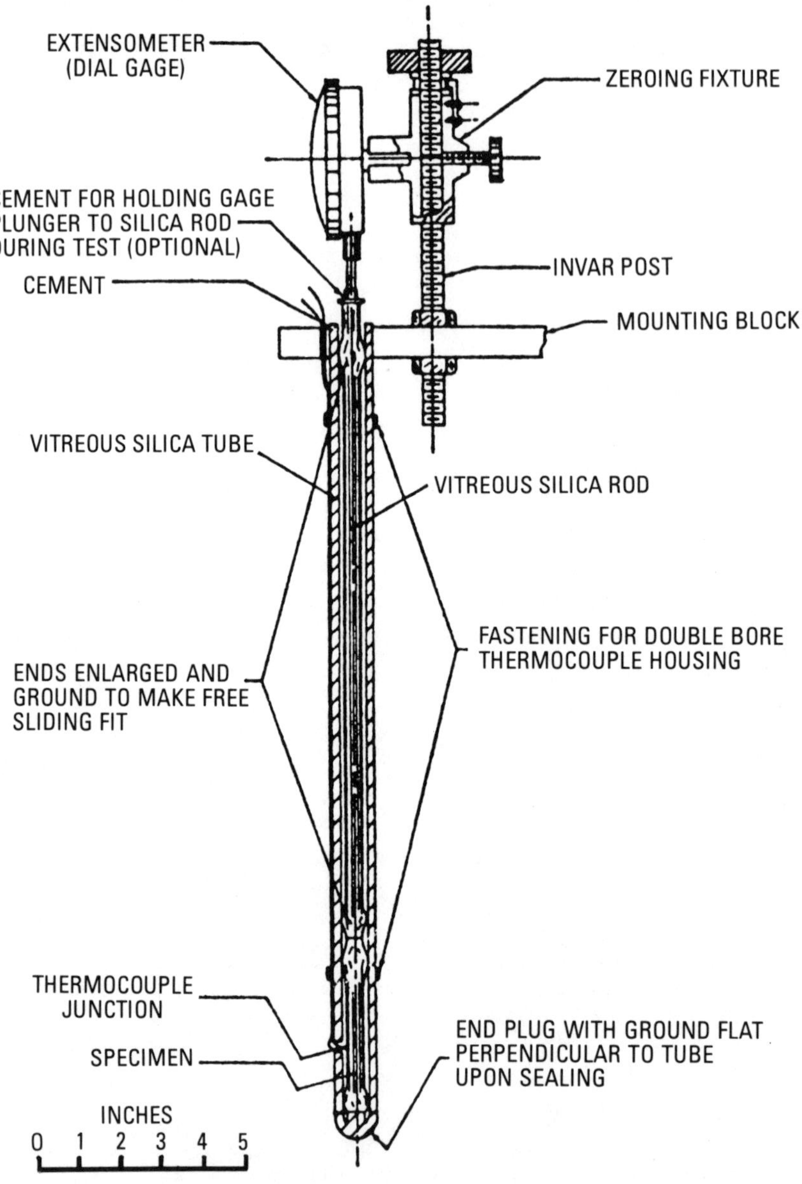

APPENDIX FIGURE 1.—Dilatometer, tube type.

ASTM E 228

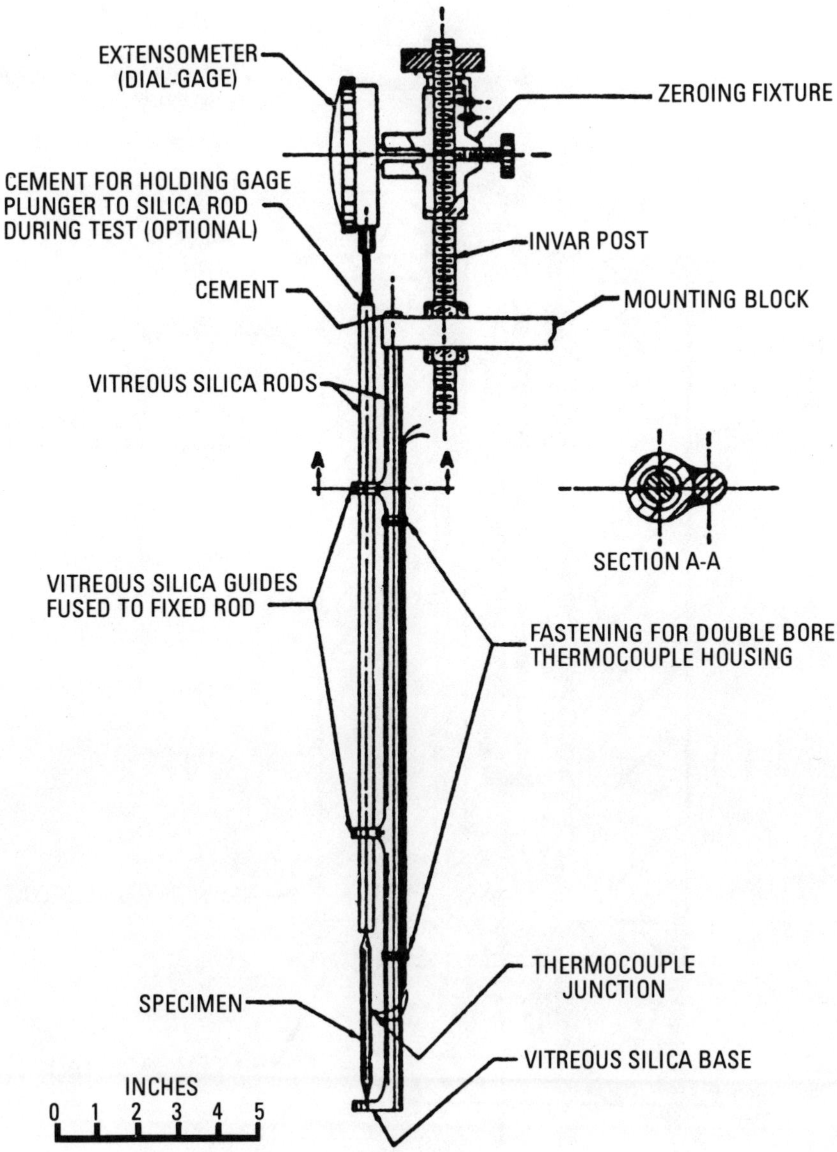

APPENDIX FIGURE 2.—Dilatometer, rod type.

E 228

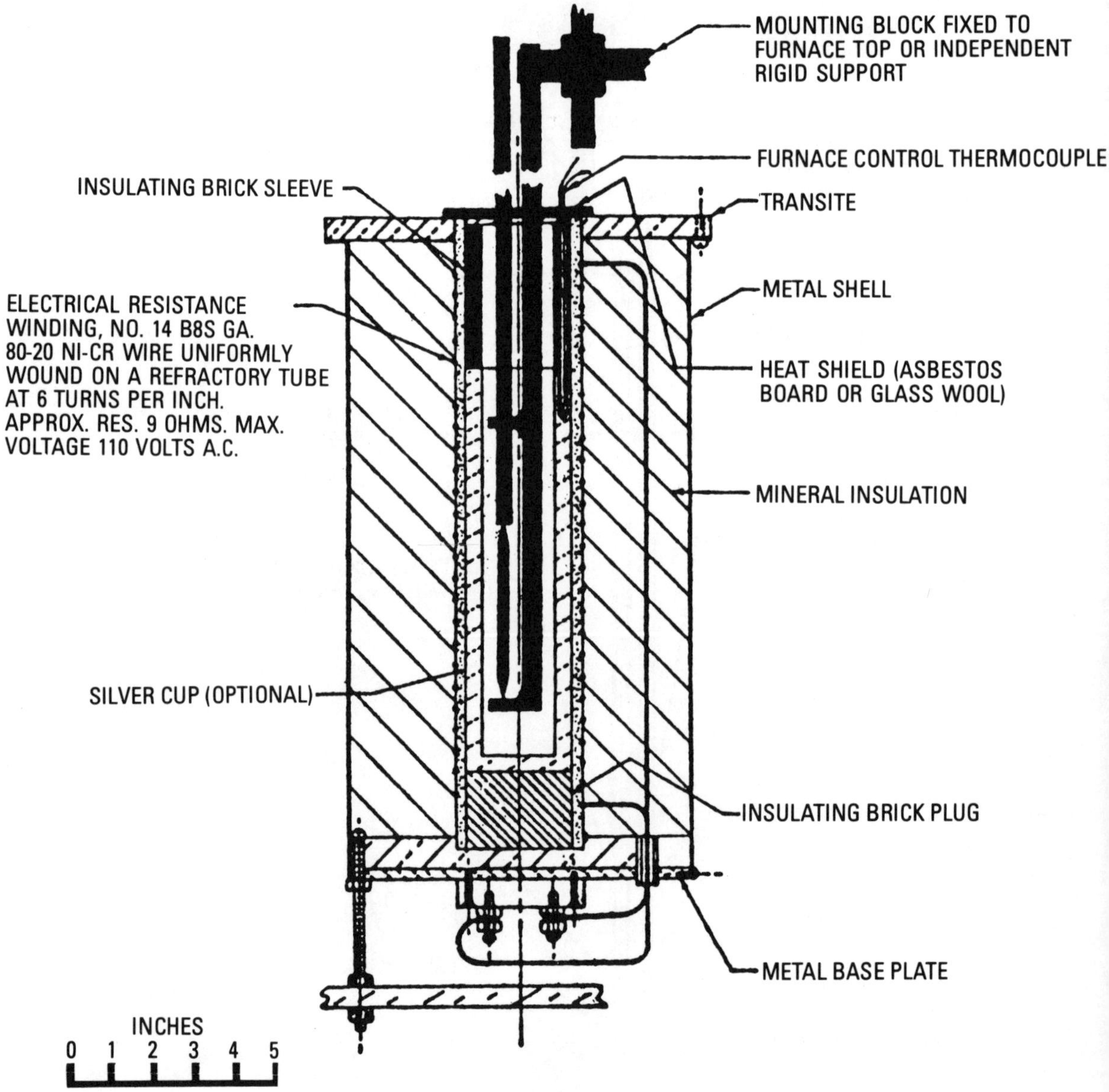

APPENDIX FIGURE 3.—Typical high-temperature furnace.

ASTM E 228

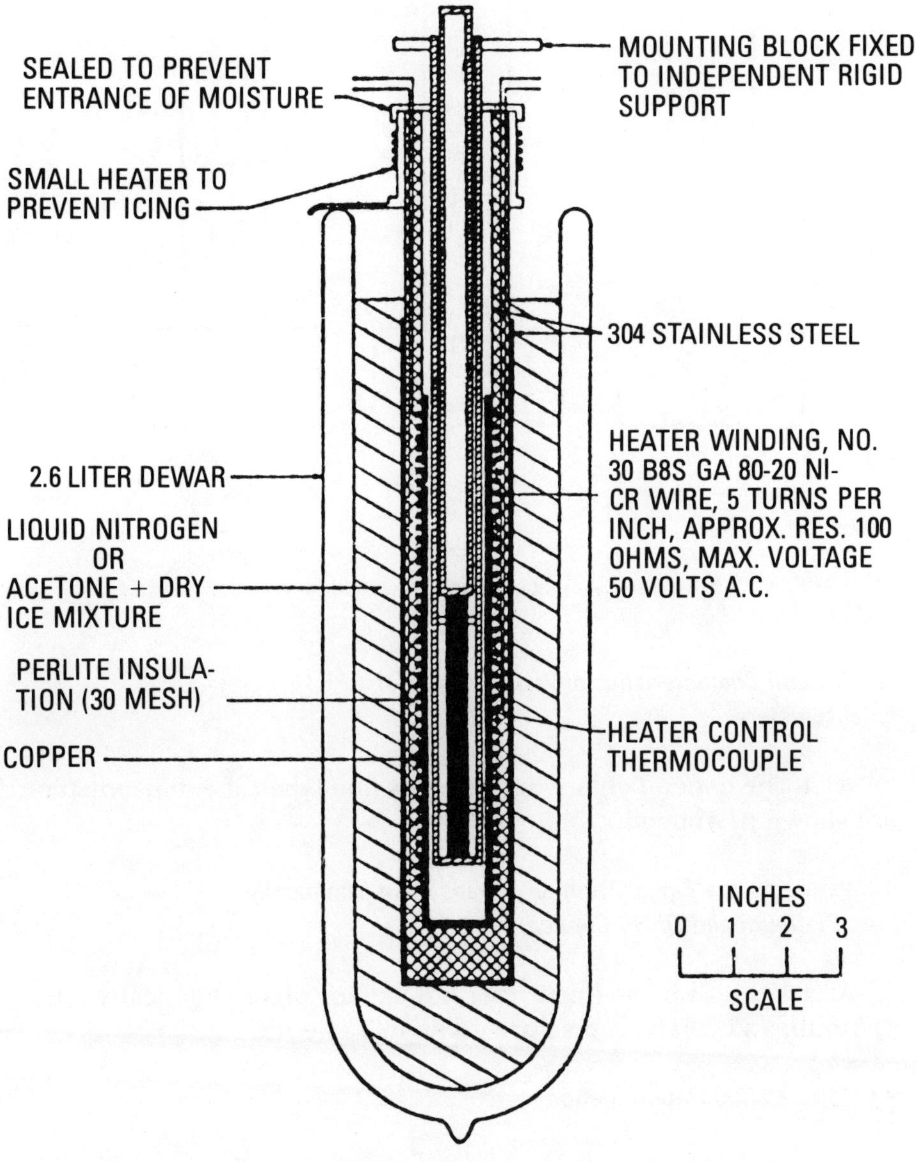

APPENDIX FIGURE 4.—Typical low-temperature furnace.

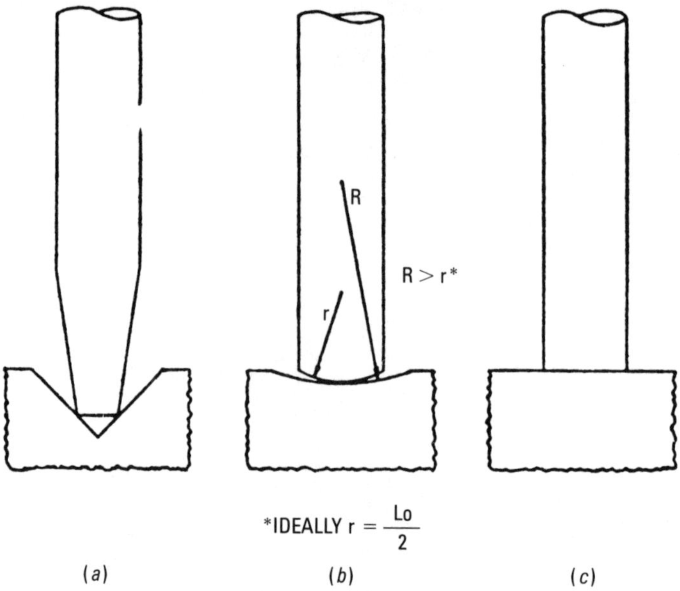

APPENDIX FIGURE–5.—Shapes of specimen ends and contact surfaces.

A1. General Characteristics of Various Types of Extensometers

A1.1 The general characteristics of some types of extensometers are shown in Appendix Table A1.

A2. Values for the Linear Thermal Expansion of Chemically Pure Platinum and OFHC Copper

A2.1 Values for the linear thermal expansion of chemically pure platinum and OFHC copper are given in Table A2.

A3. Temperature Determination

A3.1 If the thermocouple junction is embedded properly in the specimen, reliable temperatures are usually obtained under conditions of heating or cooling. However, when the junction is placed on or near the specimen surface, temperature measurements may be erroneously high on heating and low on cooling. The magnitude of this thermocouple lead or lag usually is not great but will be dependent on the distance of the thermocouple from specimen surface, specimen size, and the emissivity and thermal diffusivity of the material.

A3.2 The thermocouple lead and lag may be determined in the following manner: If the specimen to be measured is known to be reversible in thermal expansion behavior, the dilatometer assembly is heated at a constant rate over the temperature range of interest and then cooled through this region at the same rate. A plot of extensometer readings versus apparent temperature will show a temperature displacement between the heating and cooling data if there is significant thermocouple lead and lag. One half of this displacement is the lead and lag correction. This correction is then subtracted from apparent temperatures on heating and added on cooling. The magnitude will depend, of course, on the heating and cooling rate used. If the specimen is not reversible in thermal expansion behavior, another material of similar emissivity and diffusivity and known to be reversible may be substituted to determine the approximate lead and lag.

A4. Verification of Dilatometer for Errors Due to Head Assembly Temperature Changes

A4.1 Dilatometer head assembly parts, including the extensometer, which have differential motion relative to each other, should have these opposing parts constructed from low expansion materials such as invar. Regardless of materials used, the dilatometer should be verified for measurement errors due to changes in head assembly temperatures during test. The following procedure provides a means of determining the magnitude of such errors and a means of correction when needed:

A4.1.1 Determine the temperature T_L, which is the lowest ambient or head assembly temperature from which an expansion test run may be started.

A4.1.2 Determine the temperature T_A, which is the highest ambient temperature which may be encountered during an expansion test run.

A4.1.3 Determine the temperature T_H, which is the highest temperature the dilatometer head assembly reaches during an expansion test run. Determine this temperature as follows: Establish and hold the ambient temperature at T_A, then make a complete expansion test run which includes the slowest heating rate, the highest specimen temperature and the longest high-temperature hold period normally encountered. During this test run, by use of embedded thermocouples or surface contact thermometers, determine the highest temperature T_H reached by the head assembly parts.

A4.1.4 Determine the maximum head assembly temperature change ΔT as follows:

$$\Delta T = T_H - T_L$$

A4.1.5 Place a small-wattage heater (lamp bulb) near the center of the dilatometer head assembly. Wrap a loose blanket of shiny aluminum foil or paper around the head assembly and its immediate supports. Tape the foil edges to eliminate circulation of external air currents into the foil enclosure.

A4.1.6 With any specimen correctly mounted in the dilatometer, an extensometer reading is made after the dilatometer has been immersed in a controlled-liquid bath for 20 min.

A4.1.7 Measure the head assembly temperature. Turn on the heater until the head assembly reaches a temperature ΔT higher than the initial temperature. Again read the extensometer.

A4.1.8 Take the difference between extensometer readings of A4.1.6 and A4.1.7. A change of 0.00005 in. (0.0013 mm) in extensometer readings shows than an uncontrolled measurement error exists due to a change in head assembly temperatures during a test run. Uncontrolled measurement errors of 0.00005 in. or larger, which are due to head assembly temperature changes, shall be brought under control. Unless this can be accomplished by other means, the entire head assembly shall be temperature controlled. An isothermal box which contains a small fan, a heater, and thermostatic control can maintain the head at a constant temperature. If needed, a water-cooled radiation shield should also be placed between the furnace and the isothermal box. When dial gages are used, one side of the isothermal box can be made of methyl methacrylate or other transparent material.

Appendix Table A1
General Characteristics of Various Types of Extensometers

Measuring Principle	Approximate Sensitivity, in.	Range	Stability with Time
Interferometer	1×10^{-6}	long	good
Optical lever	4×10^{-6}	long	good
Unbonded wire strain gage	5×10^{-6}	long	good
Linear variable differential transformer	5×10^{-6}	long	good
Capacity micrometer	1×10^{-5}	short	poor
Magnetic gage	1×10^{-5}	short	poor
Rotating mirror	1×10^{-5}	long	good
Dial gage	1×10^{-4}	long	good
Mechanical lever	1×10^{-3}	long	good
Filar eyepiece microscope	1×10^{-5}	long	good
Electrical contacting micrometer	1×10^{-5}	long	good

Appendix Table A2
Percent Linear Thermal Expansion

TEMPERATURE, DEG C	LINEAR THERMAL EXPANSION, PERCENT	
	Platinum	Copper
−195	−0.1559	−0.2696
−150	−0.1256	−0.2228
−100	−0.0864	−0.1555
−50	−0.0442	−0.0801
0	0.0000	0.0000
25	0.0222	0.0414
50	0.0446	0.0836
100	0.0900	0.1699
200	0.1829	
300	0.2783	
400	0.3762	
500	0.4766	
600	0.5797	
700	0.6854	
800	0.7938	
900	0.9048	
1000	1.0186	

The American Society for Testing and Materials takes no position respecting the validity of any patent rights asserted in connection with any item mentioned in this standard. Users of this standard are expressly advised that determination of the validity of any such patent rights, and the risk of infringement of such rights, is entirely their own responsibility.

This standard is subject to revision at any time by the responsible technical committee and must be reviewed every five years and if not revised, either reapproved or withdrawn. Your comments are invited either for revision of this standard or for additional standards and should be addressed to ASTM Headquarters. Your comments will receive careful consideration at a meeting of the responsible technical committee, which you may attend. If you feel that your comments have not received a fair hearing you should make your views known to the ASTM Committee on Standards, 1916 Race St., Philadelphia, Pa. 19103, which will schedule a further hearing regarding your comments. Failing satisfaction there, you may appeal to the ASTM Board of Directors.

Index